TECHNIQUES OF ELECTRON MICROSCOPY, DIFFRACTION, AND MICROPROBE ANALYSIS

Presented at the
Sixty-sixth Annual Meeting
AMERICAN SOCIETY FOR TESTING AND MATERIALS
Atlantic City, N. J., June 26, 1963

Reg. U. S. Pat. Off.

ASTM Special Technical Publication No. 372

Price $3.75; to Members $2.60

Published by the
AMERICAN SOCIETY FOR TESTING AND MATERIALS
1916 Race St., Philadelphia 3, Pa.

© by American Society for Testing and Materials 1964

TN
690
.S896
1963

Printed in Baltimore, Md
March, 1965

FOREWORD

The papers in this volume were presented at a Symposium on Electron Metallography, sponsored by Subcommittee XI on Electron Microstructure of Metals of ASTM Committee E-4 on Metallography, which was held on June 26, 1963, during the Sixty-sixth Annual Meeting of the Society.

The conference chairman was J. R. Mihalisin, of the International Nickel Co. W. R. Lasko, of Pratt and Whitney Aircraft Division of United Aircraft Corp., presided over the technical sessions.

NOTE—The Society is not responsible, as a body, for the statements and opinions advanced in this publication.

CONTENTS

	PAGE
Introduction	1
Specimen Damage During Cutting and Grinding—A. Szirmae and R. M. Fisher	3
Electron Microscopy of Stress-Corrosion Cracking in AM 350 Steel for Supersonic Transport—N. A. Tiner	10
Application of a Conducting Mask for Thinning Metallic Foils for Electron Transmission Microscopy—Thomas A. Despres	24
Investigation of Microstructure and Room-Temperature Fracture in SM-200 Nickel-Base Alloy—R. W. Hertzberg and J. A. Ford	31
Microconstituents in High-Speed Steel—P. K. Koh and H. Nikkel	44
Variable Bias Illumination Control for the Electron Microscope—J. O. McPartland and J. L. Daniel	51
Two New Indexes to the Powder Diffraction File—W. C. Bigelow and J. V. Smith	54

RELATED ASTM PUBLICATIONS

Advances in Techniques in Electron Metallography, STP 339, 1962.
Electron Metallography, STP 262, 1959.
Advances in Electron Metallography and Electron-Probe Microanalysis, STP 317, 1962.

TECHNIQUES OF ELECTRON MICROSCOPY, DIFFRACTION, AND MICROPROBE ANALYSIS

INTRODUCTION

This book includes papers presented at the Technical Session on Electron Metallography held on Wednesday, June 26, 1963, during the Sixty-sixth Annual Meeting of the Society in Atlantic City, New Jersey. The session was sponsored by Subcommittee XI on Electron Microstructure of Metals of ASTM Committee E-4 on Metallography.

The papers presented at this session include new techniques and applications of electron microscopy, diffraction, and microprobe analysis in research studies.

The first paper by Hertzberg and Ford is an example of the application of electron microscopy to the study of fracture in a high-temperature alloy. The paper by Despres concerns a new technique for processing thin metal foils for viewing in transmission. Koh and Nikkel demonstrate the usefulness of probe analysis for identifying constituents in high-speed steel. Szirmae and Fisher present a critical analysis of spark cutting as a means of preliminary thinning for processing thin sections from bulk specimens.

The paper by Tiner is another example of electron fractography applied, in this case, to a corrosion problem.

Included in this volume are two papers authored by McPartland and Daniel, and Bigelow and Smith which were not presented at the technical session but, because of their recent availability and relevancy, are published here.

The first of these describes a modification for the electron microscope to provide for variable illumination. The second paper by Bigelow and Smith stems, in part, from the Subcommittee's efforts to assess the usefulness of the ASTM powder diffraction card file for electron diffraction. Two new systems are described, the Fink and the Matthews Index. Both systems are applicable to both electron and X-ray diffraction patterns.

Dr. J. R. Mihalisin acted as chairman of the Subcommittee meetings with W. R. Lasko as secretary. W. R. Lasko presided over the technical session.

SPECIMEN DAMAGE DURING CUTTING AND GRINDING

By A. Szirmae[1] and R. M. Fisher[1]

Synopsis

A detailed study has been made of the extent of deformation damage during mechanical and electrical spark cutting and planing of $3\frac{1}{4}$ per cent silicon iron during initial preparation of thin foil specimens. The results show the relation between rate of metal removal and depth of damaged layer, and demonstrate that under optimum conditions this layer can be as small as 5μ.

Due to the nature of the material or the scope of the research problem, it is sometimes necessary to prepare thin foils suitable for transmission electron microscopy from what is often referred to as a bulk specimen, in contrast to the more usual thin sheets of 2 to 6 mils (50 to 150 μ) in thickness. Since, in many cases, electro-thinning proceeds at rates as low as 1 mil (25 μ) per hr, it is impractical to carry out the final thinning on specimens of greater thickness, especially since the quality of the finished foil is usually improved when the initial thickness is very thin.

Specimens up to 40 mils (1000 μ) thick can sometimes be reduced to the desired level for electro-thinning by using chemical polishing or jet machining techniques, as described by Kelly and Nutting,[2] but beyond this thickness it is necessary to use some cutting or grinding technique as a preliminary step. Such methods produce a certain amount of deformation damage which may obscure or change features in the microstructure. Thus it is useful to know the extent of damage produced by various cutting methods commonly available in a metallographic laboratory so that the most suitable procedure can be selected. This report presents the results of measurements by etch-pitting techniques of the depth of the deformed zone resulting from mechanical cutting and grinding and spark machining of 3.25 per cent silicon iron.

Experimental Procedure

Specimens about 1 in. square and $\frac{1}{8}$ in. thick were cut from a vacuum-melted alloy containing 3.25 per cent silicon, 0.009 per cent carbon, and less than 0.001 per cent nitrogen. They were annealed for $1\frac{1}{2}$ hr in evacuated quartz capsules at 1150 C and cooled to room temperature. This treatment resulted in quite a low dislocation density and a hardness of about 200 DPH. The specimens were then cut or ground by the various methods discussed here which might be used in normal specimen preparations. They were then aged at 200 C for 2 hr to decorate the dislocations with

---

[1] Assistant scientist and staff scientist, respectively, Edgar C. Bain Laboratory for Fundamental Research, United States Steel Corporation Research Center, Monroeville, Pa.

[2] P. M. Kelly and J. Nutting, "Techniques for the Direct Examination of Metals by Transmission in the Electron Microscope," *Journal of the Institute of Metals*, Vol. 87, 1959, p. 385.

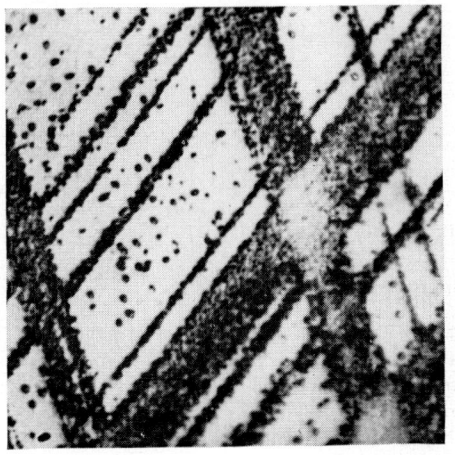 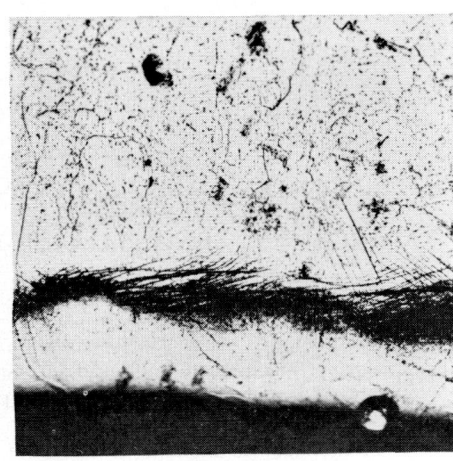

 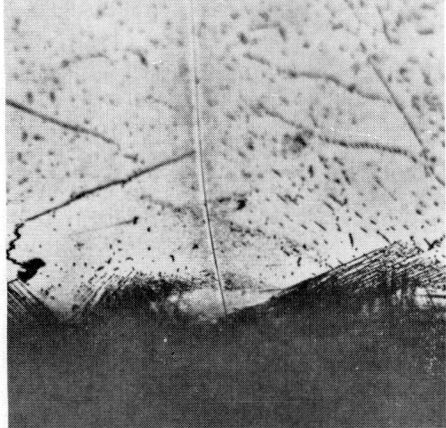

FIG. 1 (*top*)—Intersecting Dislocation Slip Bands in Bent Specimen of 3¼ Per Cent Silicon Iron (× 1000).

FIG. 3 (*bottom*)—Deformation Damage Produced by Manual Cut-Off Wheel (× 50).

FIG. 2 (*top*)—Deformation Damage Resulting From Hack-Saw Cut (× 200).

FIG. 4 (*bottom*)—Deformation Damage Produced by Precision Wafering Machine (× 300).

carbon. The decorated dislocations can readily be revealed using the chromic acid etch developed by Morris.[3] This solution is made up of 133 ml of glacial acetic acid, 25 g of chromium trioxide, and 7 ml of distilled water.

[3] C. E. Morris, "Electropolishing of Steel in Chrome-Acetic Electrolyte," *Metals Progress*, Vol. 56, 1949, p. 696.

The procedure used in this work was first to electro-polish the specimens in the chromic acid solution in a stainless steel beaker cooled to maintain the electrolyte temperature between 10 and 14 C, using 20 v and a current density of 0.3 amps/cm². After polishing for approximately ½ hr, the specimens were etched for about 5 min by reducing the voltage to 8 and the current density to 0.1 amp/cm^2. The extent of the deformation was examined by optical metallographic techniques, and the thickness of the damaged zone was measured at a number of places using a calibrated stage on an optical microscope.

that the enhanced attack or pitting occurs at the emergent dislocation sites. This explanation appears to fit in well with the observations that etch pitting does not occur in silicon irons containing less than about 2.5 per cent silicon and that the carbon segregation must exceed

TABLE 1—DEFORMATION DAMAGE DURING CUTTING OF 3¼ PER CENT SILICON IRON.

Method	Cutting Rate, mm/min	Tool Thickness, in.	Width of Cut, in.	Depth of Damage, μ	
				Average	Maximum
Hack saw	...	0.037	0.047	200	320
Jeweler's saw	...	0.008	0.012	40	60
Manual cut-off wheel, thick	5	0.059	0.061	200	1000
Manual cut-off wheel, thin	5	0.040	0.042	200	550
Precision wafering machine	10.4	0.042	0.043	25	50
	0.41	0.042	0.043	20	50
Spark-cutting machine					
(3) coarse	0.25	0.012	0.031–0.025	30	40
(5) medium	0.05	0.012	0.024–0.022	15	20
(7) fine	0.004	0.012	0.023–0.017	7	10
Spark cutting machine modified—					
Basinski very fine	0.010	0.002	0.005	0	0

TABLE 2—DEFORMATION DAMAGE DURING GRINDING AND PLANING.

Method	Surface Roughness Depth of Pits or Scratches in Microns	Rate of Removal, μ/min	Depth of Damage, μ	
			Average	Maximum
Surface grinding, coarse	10	25	80	200
Surface grinding, fine	1	0.5	25	50
Circular metallographic wet paper grinding wheel				
60 Grit	5	75	25	50
500 Grit	1	5	20	25
Circular metallographic micro-polishing cloth				
wheel, Linde C	...	0.1	3	5
Spark planing				
coarse 1	50	30	100	150
medium coarse 3	25	4	50	80
medium fine 5	15	1.3	15	25
fine 7	7–8	0.3	2	8

According to Sestak,[4] the carbon which segregates to the dislocations during the decoration heat treatment counteracts the passivity which otherwise is developed on the surface of 3.25 per cent silicon iron during electro-polishing so

some minimum amount before dislocation sites are revealed. This latter effect is illustrated in Fig. 1 which shows at 1000 × the dislocation slip band structure in a specimen deformed slightly by bending. The dislocation density in the regions where two slip bands intersect is very high, and the amount of carbon available in insufficient to decorate them to the point that they are

---

[4] Bohdan Sestak, "On the Mechanism of Rendering Visible Dislocations on the Surface of Iron Crystals by Anodic Dissolving," *Czechosl. Journal of Physics*, Vol. 9, 1959, p. 339.

revealed by etch-pitting. This effect was encountered very frequently in the investigation. As the micrograph shows, the individual dislocations are only resolved in the very narrow slip bands and in the areas between them. No attempt was made to measure the dislocation density in the deformed regions but only the depth of deformation beneath the cut surface.

RESULTS

The appearance and depth of the deformed regions were found to vary considerably, depending on the cutting or grinding method used. Optical micrographs illustrating several different examples are shown in Figs. 1 to 8. The results of the measurements are summarized in Table 1 for deformation damage resulting from cutting and in Table 2 from grinding and planing. The average values given in the tables refer to the limit of the rather uniform zone of etch pits, whereas the maximum refers to unusually deep penetration observed at a few points during a traverse of several millimeters along the deformed edge.

Cutting:

The micrograph in Fig. 2 shows the appearance of a manual hack-saw cut. It is seen that the heavily deformed regions along the edge are not etch-pitted as discussed above. The rate of cutting is not presented in Table 1 because it obviously depends on the energy of the operator and other factors, but in this case, with a specimen thickness of $\frac{1}{8}$ in., it was about 2 mm/min. As shown in Table 1, the damage produced by manual saw cutting is considerably less if a jeweller's saw is used.

Most metallographic laboratories have a manual-feed abrasive cut-off wheel available for use. Figure 3 shows the damage produced using the standard 0.059-in.-thick wheel. Table 1 shows that the maximum damage is very much greater than the average, probably due to chatter caused by uneven specimen feed. The use of a somewhat thinner wheel reduced the depth of the deepest regions, but the average was unchanged.

Abrasive cutting, using a high speed, precision wafering machine with an automatic specimen feed, was found to produce considerably less damage than the manual cut-off wheel. A typical deformed area is shown in Fig. 4. As indicated in Table 1, the layer is only $\frac{1}{10}$ as thick as for the manual type cut-off wheel, and the use of a very slow rate of automatic specimen advance did not reduce the damage to any extent.

A relatively new method of cutting is spark-machining, where an electric spark discharge is maintained between the specimen and the tool electrode. This causes rapid erosion of both surfaces. The intensity of the spark is controlled by varying the amount of capacitance in the discharge circuit, which acts in conjunction with a servo-mechanism to maintain the correct spacing between tool and specimen. The shape of the cut is the same as that of the tool, and therefore holes of any shape can be produced. The specimen and cutting tool are usually immersed in parrafin oil. A commercial spark erosion machine known as the Servo-met was used in this investigation.

The data presented in Table 1 show that the damage is reduced when slower cutting rates are used, but it is clear that it is never negligible, as is sometimes assumed. The appearance of the deformed regions after cutting is quite similar to that produced during spark planing, which is described in the next section.

Some sparking occurs between the

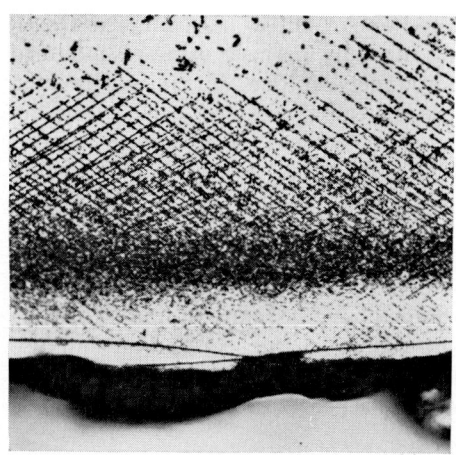

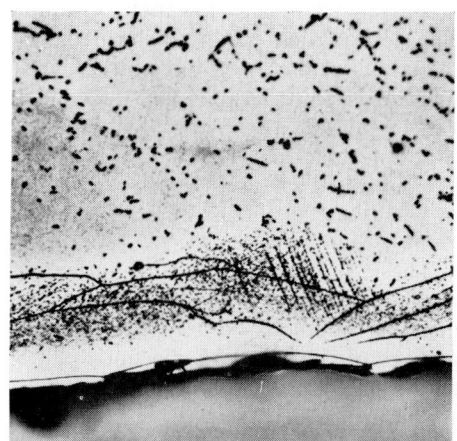
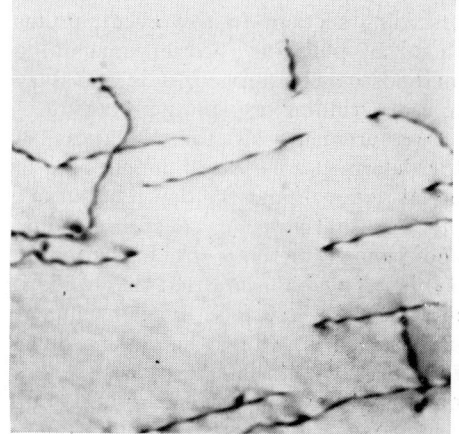

FIG. 5 (*top*)—Deformation Damage Produced by Metallographic Grinding on 500 Grit Paper (× 500).

FIG. 7 (*bottom*)—Deformation Damage Produced by Medium Spark Planing (× 500).

FIG. 6 (*top*)—Deformation Damage Produced by Coarse Spark Planing (× 500).

FIG. 8 (*bottom*)—Dislocation Sub-Structure in an Annealed, $3\frac{1}{4}$ Per Cent Silicon Iron Specimen Ground to 1 Mil Thickness Prior to Final Thinning (× 60,000).

tool and the sides of the cut region in the specimen resulting in a tapered cut and also in a reduction in the cutting rate. This difficulty, as well as problems with bending of thin slices toward the tool as the cut nears completion due to electrostatic attraction, can be eliminated by using a fine wire 0.002 to 0.004 in. in diameter as a cutting tool which is continuously unreeled and drawn over the specimen. Basinski[5] has modified a Servo-met machine to use a fine wire in combination with a reduced spark intensity to decrease specimen damage without any loss in cutting

---

[5] Z. S. Basinski, private communication, 1963.

rate. The result of measurements of a cut made by Basinski using his modified Servo-met on one of the silicon iron specimens prepared for this study are listed in Table 1. The damage in this case is less than the limit of measurement, which is about 2 μ.

The origin of the deformation damage produced by spark machining is not well understood but is probably a result of thermal stresses due to rapid heating and cooling at the spark crater and also from shock-waves produced by the spark discharges.

Grinding-Planing:

In the case of bulk specimens less than $\frac{1}{8}$ in. thick, it is not feasible to use any of the cutting methods described in the preceding section. In this event, unless chemical polishing or jet machining methods can be applied, it is necessary to use a grinding or planing procedure.

Measurements of the thickness of the deformed zone, the nominal rate of metal removal, and the depth of surface pits or scratches for several grinding and planing methods are presented in Table 2. The data show that the standard machine shop surface grinder can be used safely to reduce thicknesses down to about 12 mils (300 μ) if a sufficient number of fine finishing cuts are taken to remove the effects of the coarse stages. However, proper finishing is quite time-consuming, and it is difficult to properly hold even highly magnetic materials in the form of thin sheets.

The deformation caused during grinding with wet abrasive paper on a circular metallographic wheel is also listed in Table 2. Fresh, that is, unused disks were used for these measurements, and normal precautions were taken to maintain adequate cooling and avoid excessive pressure. The specimens are held, for hand grinding or polishing, on metal or bakelite blocks of convenient size by double-sided adhesive tape. Care must be taken to remove the specimen from the tape without bending it, and it is best to remove the adhesive with a solvent. Figure 5 shows an example of the deformation zone on a specimen ground with 500 grit carborundum paper. The damage resulting from polishing on a cloth wheel was found to be very slight as indicated in Table 2, but by the same token, considerable time is required to remove all the damage caused by paper grinding.

The Servo-met spark machine has a planing attachment which consists of a disk which slowly rotates above the specimen surface. It is a problem to hold very thin specimens, and sometimes it is necessary to use a low melting solder. The deformation produced during spark planing is summarized in Table 2. Some illustrative micrographs are shown in Figs. 6 and 7. These micrographs show that fine cracks were often found near the deformed surface.

DISCUSSION

These measurements of deformation damage caused by various cutting and grinding methods largely confirm the efficacy of the standard specimen preparation procedures evolved in this laboratory during the past several years. The tables of data show that the extent of damage during preparation can vary considerably between various techniques. However, equally important are the large differences in rate of metal removal, which make methods such as spark cutting and planing at the minimum spark intensity and polishing with cloth too slow to be practical for regular use.

It is obviously impossible to specify an exact margin of safety that should be used to avoid artifacts resulting from any of the preparation techniques described. When care was taken, it was found that specimens could be ground to

2 mils prior to electro-thinning without introducing dislocation sub-structure into the normal, essentially dislocation-free annealed structure. Figure 8 shows an electron micrograph of an annealed specimen ground to 1 mil prior to thinning. The resulting dislocation density is not large, but it is still higher than that usually found in an annealed specimen, and its occurrence could be misleading if its origin was not recognized.

Although these measurements were made on 3.25 per cent silicon iron, experience in the laboratory suggests that they apply to many other ferrous materials of comparable yield strength. However, for high-purity iron, the deformation damage appears to be nearly twice as deep as that recorded in these tables. Samuels[6] has made similar measurements of the depth of specimen damage during spark planing of 70-30 brass. Although he did not report rates of metal removal, his results suggest that for comparable spark intensity, the damage in brass penetrates to about twice the depths reported here.

As indicated by these observations, a very satisfactory method of preparing foils from bulk specimens is as follows:

1. Cut off thin slices about 10 to 15 mils thick with a precision wafering machining.
2. Grind with 60 grit paper to 8 mils and with 500 grit to 4 mils.
3. Finish foil preparation with a normal electro-thinning procedure.

In conclusion, with sufficient care normal metallographic cutting and grinding techniques can be used to prepare thin specimens from bulk samples suitable for final electro-thinning.

Acknowledgments:

We are indebted to Dr. Basinski for preparing a specimen on his modified spark-cutting machine and to numerous members of the Fundamental Research Laboratory for comparing notes on preparation procedures.

---

[6] L. E. Samuels, "Surface Damage Produced by the Electric-Spark Cutting Process," *Journal of the Institute of Metals*, Vol. 91, 1963, p. 191.

ELECTRON MICROSCOPY OF STRESS-CORROSION CRACKING IN AM 350 STEEL FOR A SUPERSONIC TRANSPORT

By N. A. Tiner[1]

Synopsis

Stress-corrosion behavior of AM 350 steel has been evaluated in natural atmosphere in close proximity to the oceans. A number of stress corrosion-cracked specimens have been examined by means of electron microfractography. The salient microstructural features of crack initiation and propagation are described and interpreted.

The characteristics and performance requirements of a supersonic transport have been reviewed by Raring et al (**1,2,3**).[2] It has been pointed out that the operating life of a supersonic transport must be a minimum of 32,000 hr over a twelve year period. The leading edges of the wings and certain areas near the nose of a mach 3 plane would have maximum temperatures of about 550 to 600 F. The rest of the plane will be at somewhat lower temperatures. The structure will have to be maintained in a completely sound and integral condition. Since the plane will operate in close proximity to the oceans, the skin must resist salt corrosion, and the structural integrity must not be complicated by possible stress-corrosion cracking.

The precipitation-hardening stainless steels, titanium alloys, and the nickel, cobalt, and iron base superalloys appear to be very attractive materials for the fabrication of the mach 3 supersonic transport. Screening tests on these materials are now being made by Douglas Aircraft under the technical support of USAF Aeronautical Systems Division and NASA to determine how close they come to the design properties demanded by the mach 3 aircraft.

This investigation is an attempt, in a qualitative way, to determine the mode of stress-corrosion crack initiation and propagation in the semiaustenitic precipitation-hardening stainless steel type AM 350. Specimens are sheared from sheets, heat-treated, machined into tensile coupons, cleaned, stressed on single-spring loading fixtures, and exposed to continuous salt spray and natural beach corrosion environment. The surfaces and fracture faces of the specimens that have failed have been examined by optical and electron microfractography.

During the last two decades many facts concerning the stress-corrosion cracking of austenitic and martensitic chromium-nickel steels in chloride solutions have been established. The mode of development of stress-corrosion cracks has been examined primarily by optical metallographic techniques by sectioning

---

[1] Head, Materials Research Dept., Astropower, Inc., a subsidiary of Douglas Aircraft Co., Newport Beach, Calif.

[2] The boldface numbers in parentheses refer to the list of references appended to this paper.

TABLE 1—CHEMICAL COMPOSITION OF AM 305 IN CONDITION SCT 850.

Chemical Composition, mill report

Code	Heat No.	Carbon	Silicon	Manganese	Chromium	Nickel	Molybdenum	Iron	Phosphorus	Sulfur	Nitrogen
A1	89324	0.086	0.30	0.82	16.51	4.34	2.72	Bal.	0.022	0.017	0.092
A2	79928	0.090	0.29	0.82	16.60	4.31	2.90	Bal.	0.018	0.012	0.090

TABLE 2—MECHANICAL PROPERTIES OF AM 305 IN TRANSVERSE DIRECTION AND CONDITION SCT 850.

Specimen	Heat treatment	0.2% Yield, ksi	Ultimate, ksi	Elongation, %
A1-212	basic temper	167.6	208.9	10
207	basic temper	173.5	206.3	10
203	basic temper	175.7	205.8	8
		avg 172.5	avg 207.0	avg 9.3
A1-208	basic temper + 1000 hr at 650 F	177.4	209.7	7.8
209	at 650 F	177.4	209.3	8.0
213	at 650 F	177.4	209.3	7.5
		avg 177.4	avg 209.4	avg 7.8

TABLE 3—STRESS CORROSION BEHAVIOR OF AM 305 IN CONDITION SCT 850.

Days to failure in stress corrosion tests, at El Segundo Beach, Calif.

	Heat No. 89324						Heat No. 79928				
Stress, % of yield	30	40	50	60	75	90	30	50	60	75	90
Stress, ksi	51.7	68.9	86.10	103.3	129.5	155.0	53.3	88.8	106.6	133.3	159.9
SCT 850	NF14	NF12	16	37	2	2	NF14	16	4	1	1
	NF14	12	32	10	7	2	NF14	16	1	1	5
	NF14	NF12	30	19	2	5	NF14	5	8	7	2
				16	42	6	2		5	4	2	1
				11		2	2		4	4	1	
SCT 850 plus 1000 hr at 650 F	NF12	NF12	6	14	7	4	...				
	12	12	16	6	5	4	...				
	12	12	13	8	9	4	...				
			6	9	5	4	...				
			6		5	6					

the specimen at various stages of crack development (4). Attempts were also made by Nielson et al (5) to study the fine details of the actual fracture faces produced in stainless steels by stress-corrosion cracking.

Most of the results described in this paper are concerned with the fine structural details brought about by electron microscopic observation of replicas from direct or etched fracture faces and from their profiles. It is hoped that these studies will furnish new and important information pertinent to the mechanism of the stress-corrosion cracking process.

MICROSTRUCTURAL CHARACTERISTICS AND STRESS-CORROSION BEHAVIOR

The experimental material used for the present work was from two heats of AM 350 steel. Specimens were prepared from 0.025-in.-thick sheets in transverse direction and heat treated to condition SCT 850 (short-time anneal at 1710 F, cooled to −100 F, and tempered at 850 F for 3 hr.), or reheated at 650 F for 1000 hr. Stress-corrosion tests were conducted at El Segundo Beach, Calif.,

chromium carbide ($Cr_{23}C_6$) primarily at the austenite-delta ferrite interfaces. Cooling to −100 F produces a hardness increase, possibly by a chromium-rich ferrite precipitation.

The typical electron microstructures of the specimens in condition SCT 850

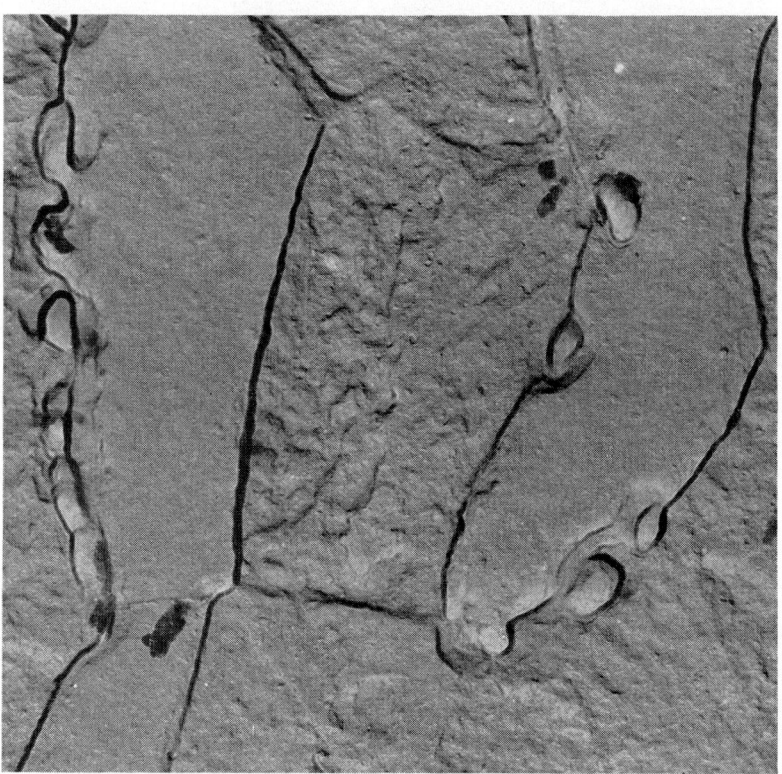

Etched with 30 per cent hydrochloric acid (HCl), 20 per cent nitric acid (HNO_3), and 50 per cent water (H_2O) solution. Positive two-stage parlodion-carbon replica shadowed with chromium.

FIG. 1—Specimen of AM 350 in Condition SCT 850 (final magnification ×24,000).

at 50, 60, 75 and 90 per cent yield strength by means of spring-loaded stress jigs. The chemical composition, mechanical properties, and the results of stress corrosion tests are presented in Tables 1, 2, and 3.

The experimental alloy, as annealed, generally contains 0.05 to 0.06 per cent carbon in solid solution, and the remainder is precipitated in the form of

and after reheating 1000 hr at 650 F are shown in Figs. 1, 2, and 3. The chromium carbide particles are clearly visible, primarily at the delta ferrite boundaries. The areas with acicular appearance in Fig. 2 are believed to be martensite plates, whereas the relatively smooth zone in the center between the two delta ferrite crystals is the retained austenite dispersed among the martensite plates.

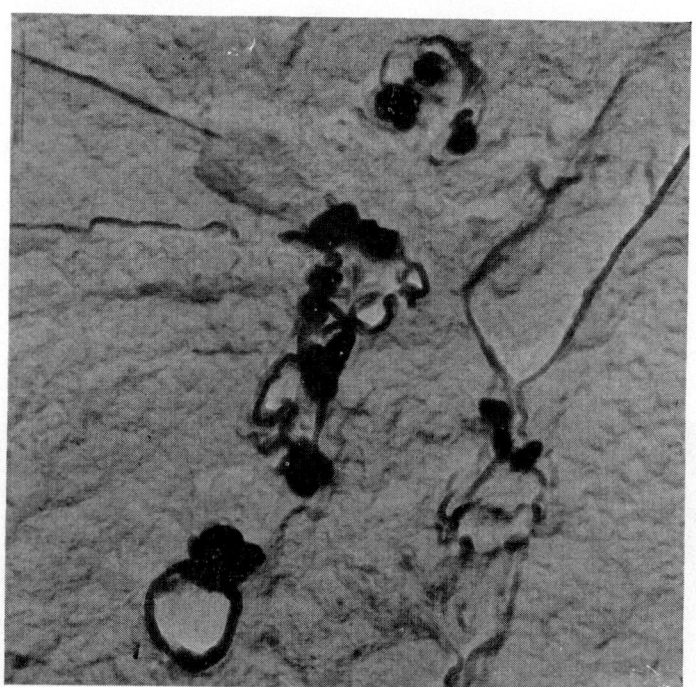

FIG. 2—Specimen of AM 350 Illustrating the Predominantly Martensitic Matrix and Some Retained Austenite Dispersed Among Martensite Plates (final magnification ×24,000).

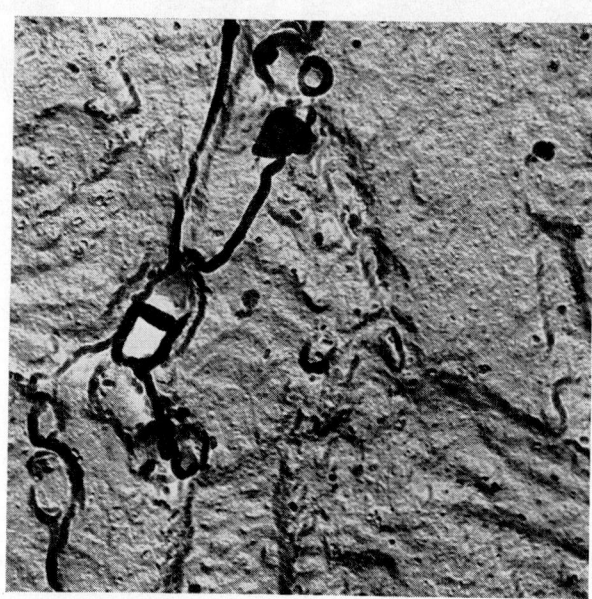

Etched with 30 per cent HCl, 20 per cent HNO_3, and 50 per cent H_2O solution. Positive two-stage parlodian-carbon replica shadowed with chromium.

FIG. 3—Specimen of AM 350 in Condition SCT Exposed to Heat at 650 F for 1000 hr (final magnification ×24,000).

The precipitate causing the hardness increase during tempering at 850 F could not be identified. No significant change in the electron microstructures was noted when the specimens were exposed to heat for 1000 hr at 650 F, although strength appeared to be slightly increased.

The results of stress-corrosion tests in natural beach atmosphere, as well as in salt spray environment, indicate that edge to the center in some specimens, or in the center portion only in the others. The brittle regions have a granular appearance and represent stress-corrosion cracked surfaces by slow crack propagation, whereas the ductile regions have a fibrous structure and represent failure by shear after the crack propagated to a certain length as shown in Fig. 4. The intermittent white streaks are in the transverse direction parallel with the

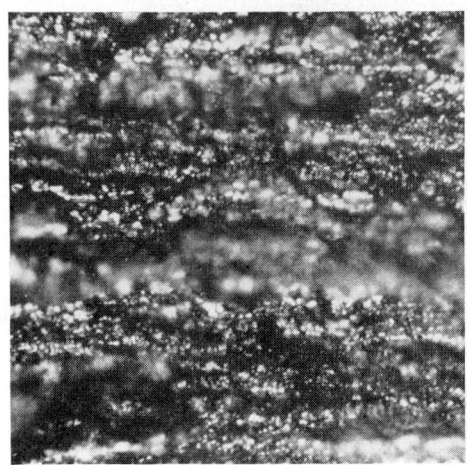 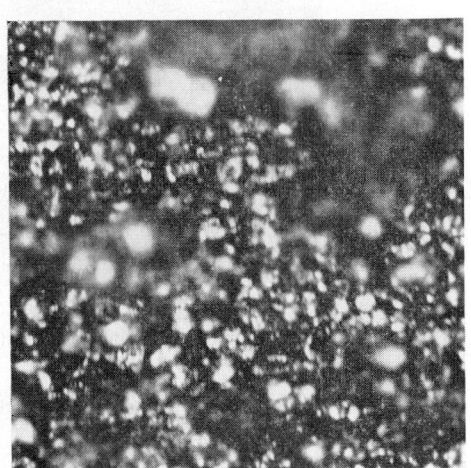

(*left*) Shear Lip Region.
(*right*) Slow crack propagation, brittle fracture region.

Fig. 4—Fracture Face of AM 350 Illustrating Failure in Stress Corrosion Test in a Natural Beach Atmosphere.

AM 350 in condition SCT is susceptible to stress-corrosion cracking. This tendency was detectable at 30 to 40 per cent yield strength (or 50 to 60 ksi), and became very severe at 80 to 90 per cent yield strength (or 130 to 150 ksi).

Optical microscopic observation of the surfaces and fracture faces showed that AM 350 specimens were susceptible to general pitting corrosion. The fracture faces of the specimens which failed in corrosion tests exhibited two distinct regions: one a ductile type and the other a brittle type whihc occurred from one sheet surface and are approximately perpendicular to the direction of the tensile load. The shear lip generally is inclined at about 45 deg to the direction of the load.

Electron Metallography of Crack Initiation

The examination by optical and electron microscopy of areas on AM 350 steel in which there has been "general" attack by natural beach environment revealed that the microcracks in the stress-corrosion specimens originate from

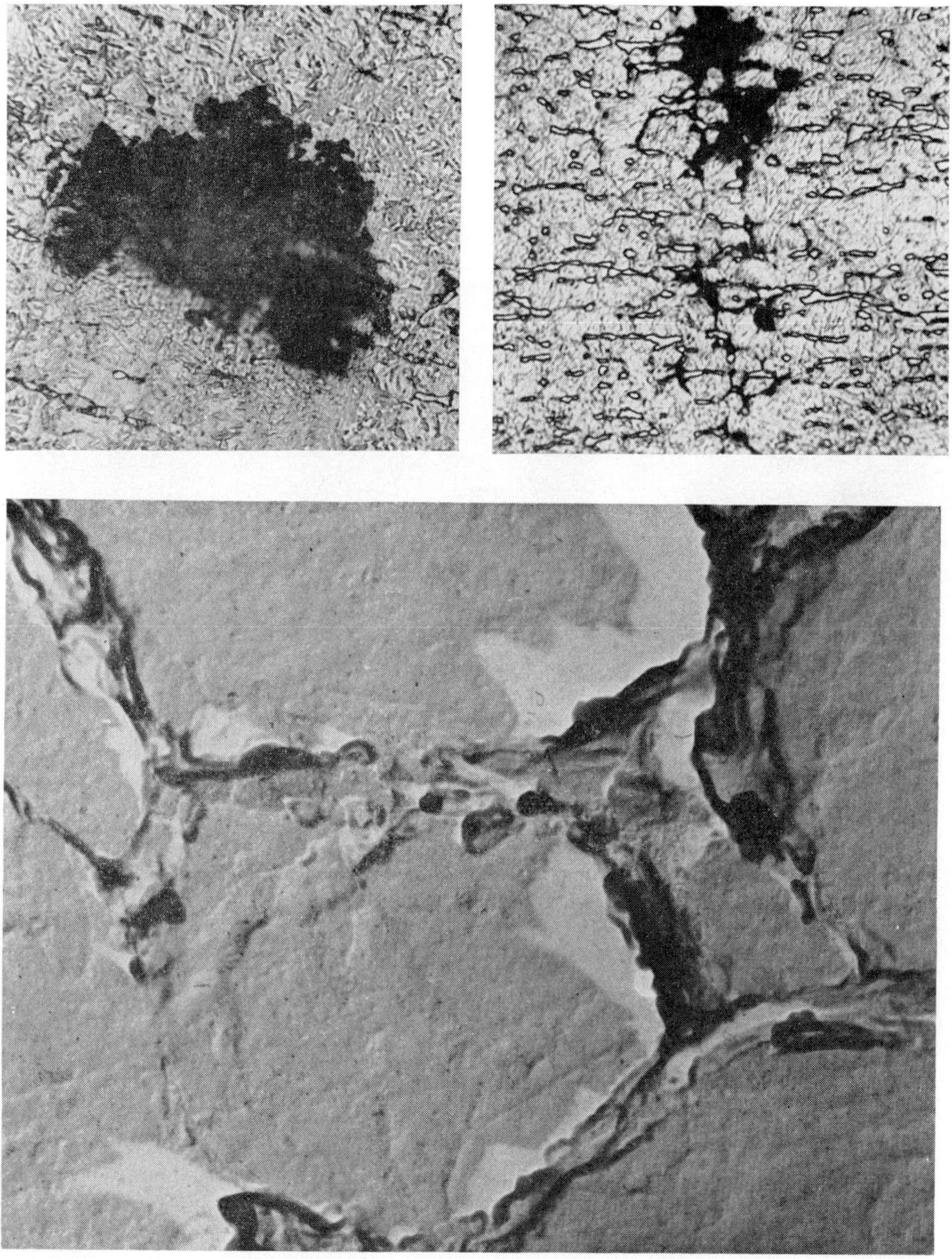

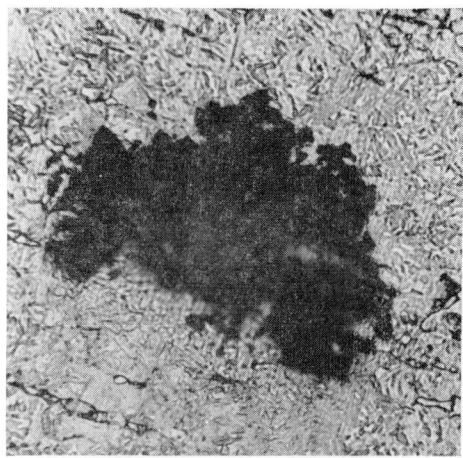
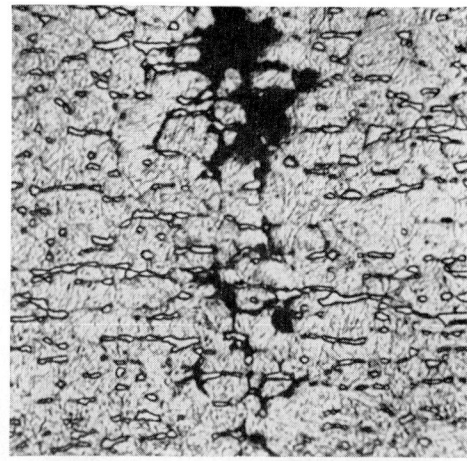

(*top left*) Slightly polished and etched surface (×200).
(*top right*) Microsection at a point bisecting the pit (×200).
(*bottom*) Electron micrograph illustrating the path of the microfissures near the tip of crack shown in above micrograph (*top right*). Etched with HCl-HNO-H_2O solution (final magnification ×16,500).

FIG. 5—Specimen of AM 350 Illustrating Pitting by General Corrosive Attack.

pits. Pitting generally occurs at random over the entire surface of a specimen and can be detected under optical microscope, as shown in Fig. 5 (*top left*). Most of the pits are shallow and saucer-like in shape. A number of pits are associated with large intergranular cracks extending inward as can be shown by cross sectioning the specimen at a point bisecting the pits, as shown in Fig. 5 (*top right*).

Replicas taken from the sheet surface near the fracture edge, in general display a mottled appearance with island-like configurations standing in relief. In certain areas at the edge of a relief zone there were small lens-shaped cracks. Electron micrograph (Fig. 6) shows fine veins connected to the comparatively large fissure, possibly a stress corrosion attack.

ELECTRON METALLOGRAPHY OF CRACK GROWTH

The mode of stress-corrosion crack propagation has been investigated by

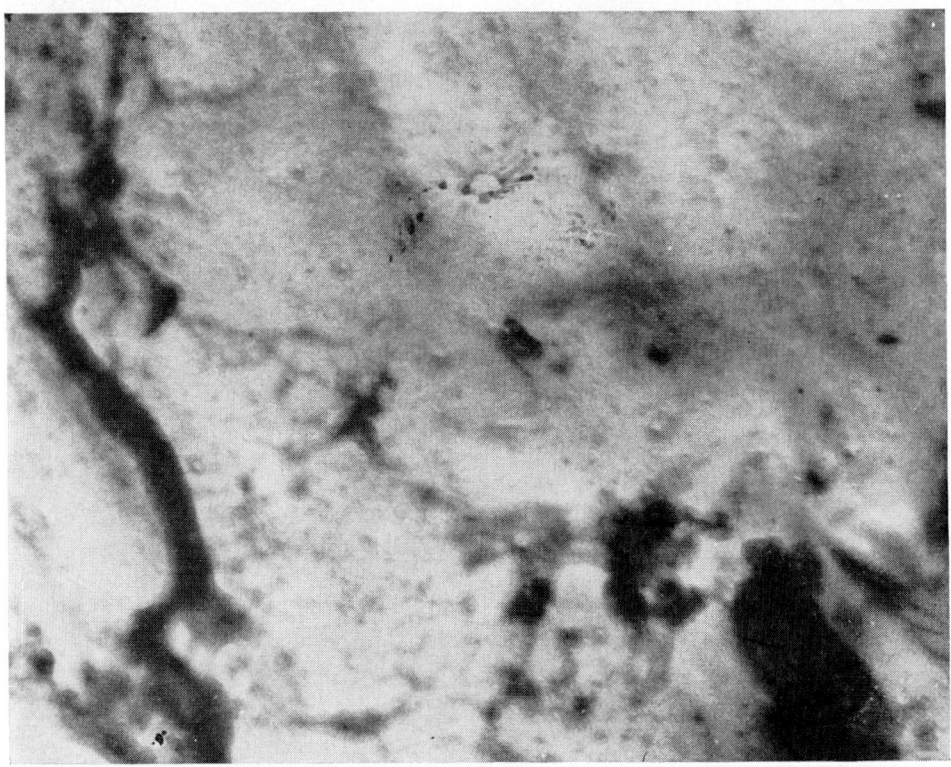

FIG. 6—Specimen of AM 350 with Positive Two-Stage Parlodian-Carbon Replica from Stress-Corrosion Surface Near Area of Failure (final magnification ×21,000).

replication of direct or etched fracture faces and their profiles. The method of replication consisted of pressing a piece of cellulose-acetate sheet, which had been softened with acetone, against the fracture face. The plastic was allowed to harden in place, then stripped from the fracture face. This negative plastic replica was shadowed with chrome at 45 deg, followed by deposition of a layer of carbon at 90 deg. The plastic was then dissolved in acetone and the carbon replica was placed in rectangular grids by carefully orienting the sheet surface parallel to the long side of the rectangles of the grid and examined in the electron microscope. The profile of fracture faces was prepared by electroless nickel-plating the fracture face, microsectioning, and replicating the polished and etched section. The salient microstructural features noted are shown in the following series of electron micrographs.

Figure 7 shows a two-stage chromium shadowed plastic-carbon replica of a specimen which failed under static stress in natural atmosphere at El Segundo Beach. The replica is taken from the ductile fracture region and illustrates the appearance of the fracture face, failed by shear after the stress-corrosion crack propagated from one edge almost to the center of the specimen. There are para-

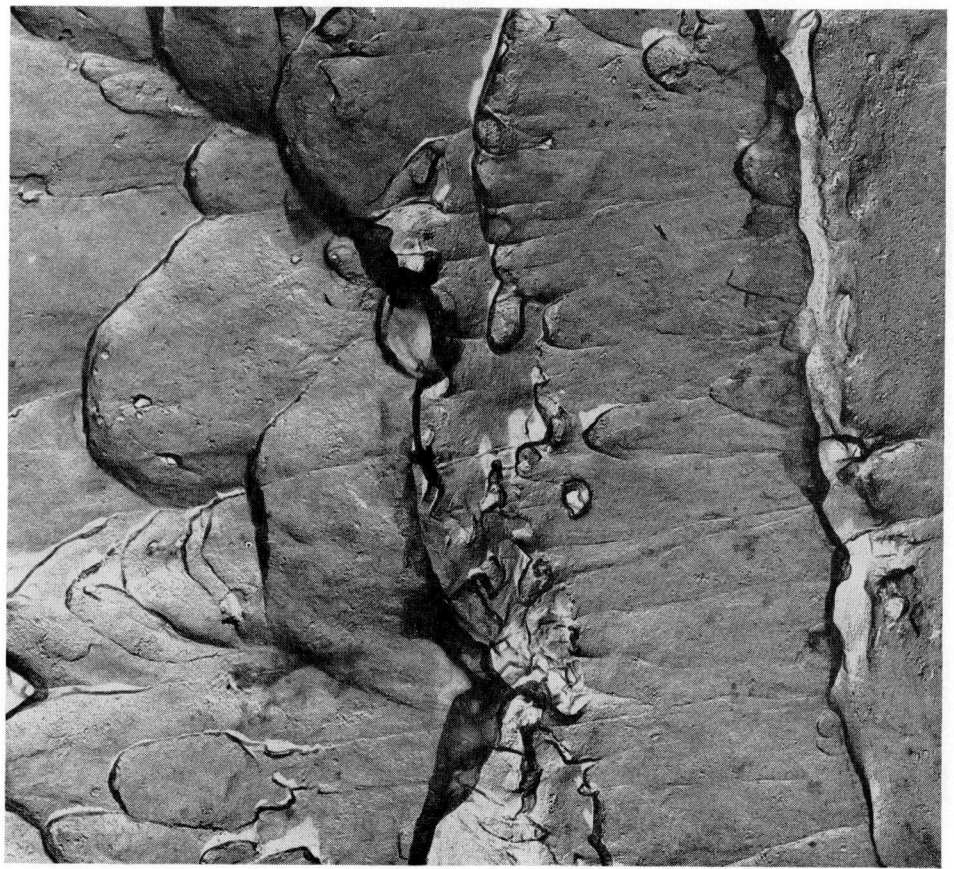

Electron micrograph illustrating the fracture face in the shear lip region. The direction of the tensile load is perpendicular to the white streak.

FIG. 7—Specimen of AM 350 Showing Failure Under Static Stress in Natural Atmosphere at El Segundo Beach (final magnification ×16,500).

bola-like lines or dimples in areas between a series of intermittent light and dark streaks or microfissures running approximately perpendicular to the direction of the tensile load and parallel to the sheet surfaces. In certain areas the parabola-like lines or dimples appeared to be absent or very few, and the streaks are prominent and continuous toward the fracture origin. The dimples become more common in occurrence toward the shear lip region. This was determined by

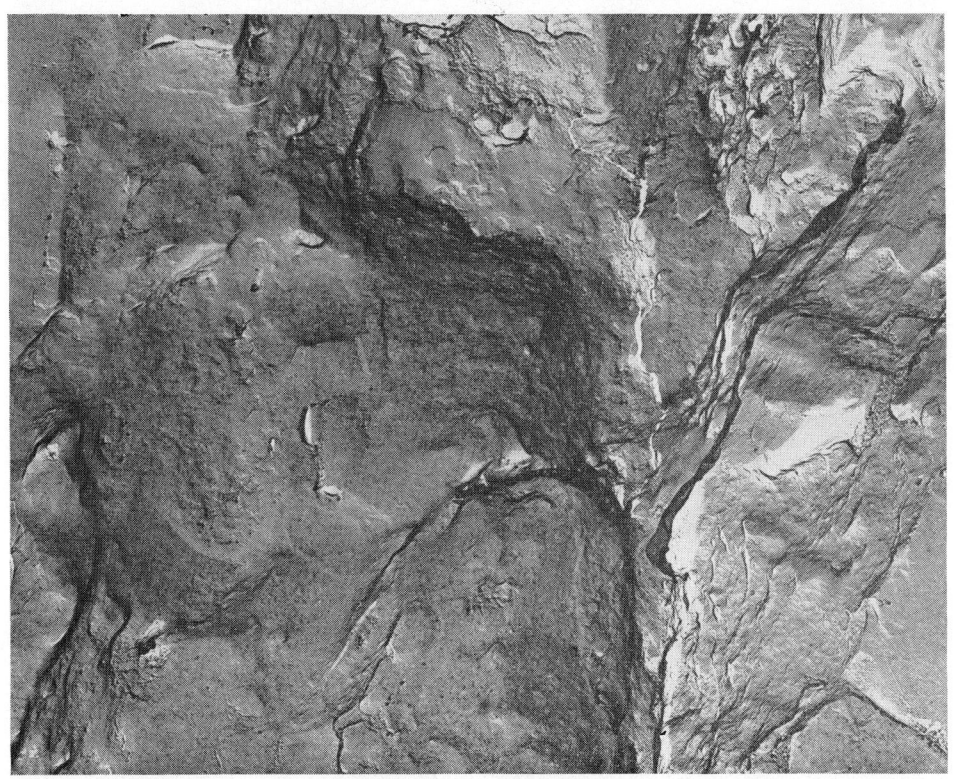

Electron micrograph illustrating the fracture face in the slow crack-propagation region. The direction of tensile load is perpendicular to the dark streak.

Fig. 8—Specimen of AM 350 Showing Failure Under Static Stress in Natural Atmosphere at El Segundo Beach (final magnification ×16,500).

dimples are numerous and the streaks are not prominent and broken up. The direction of the axis of symmetry of the parabola-like lines is in the shear plane perpendicular to the streaks.

Figures 8 and 9 show two-stage replicas taken from the brittle fracture region of specimens which failed under static stress in natural beach atmosphere. The shadow casting each replica from the direction of fracture origin and carefully examining the microstructure with this in mind. The streaks were always parallel to the direction of crack propagation and perpendicular to the direction of tensile load.

Figure 10 depicts two-stage replicas taken from the fracture faces of the specimen shown in Figs. 7 and 8 after it was etched by immersion in a solution of hydrochloric acid (HCl), nitric acid (HNO$_3$), and water (H$_2$O). The direction of crack propagation is parallel with the streaks. The etchant used, in general, creates pits with sides parallel to the cubic planes of the ferrite crystals. Thus the traces of the etch pits can be used to determine the direction followed by the fracture (6). It can be stated that the flat zones were delta ferrite, the streaks were the boundaries of delta ferrite, and that the delta ferrite have a preferred orientation, the cube diagonal of crystal lattices being parallel to the rolling direction, approximately the same as the direction of the streaks.

Figure 10 (*center*) clearly shows that the streaks or microfissures were indeed the boundaries of delta ferrite. Along these boundaries and also within the presumably martensite matrix, carbide

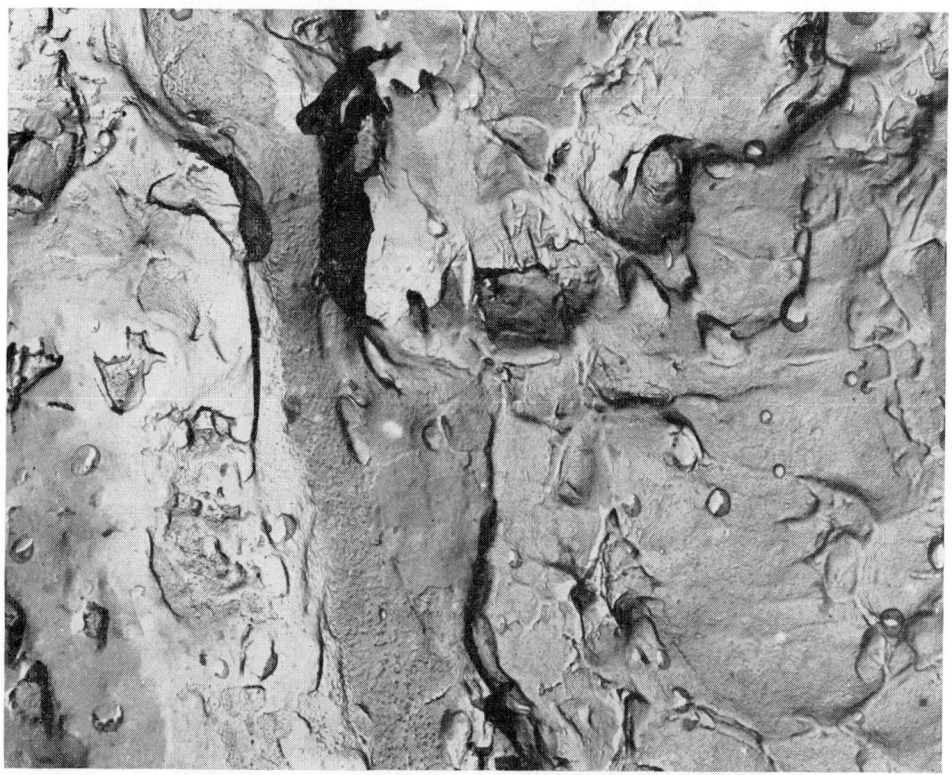

Electron micrograph illustrating the fracture face in the slow crack-propagation region near the shear lip. The direction of the tensile load is perpendicular to the dark streaks.

FIG. 9—Specimen of AM 350 Showing Failure Under Static Stress in Natural Atmosphere at El Segundo Beach (final magnification ×18,000).

particles can be identified. White spheroids are probably sites of carbides that were torn out during crack propagation or etching. Spheroids associated with shadows are actual carbides brought out in relief by etching. Figure 10 (*bottom*) shows the appearance of dimples in the etched areas near the shear lip of the stress-corrosion cracked region.

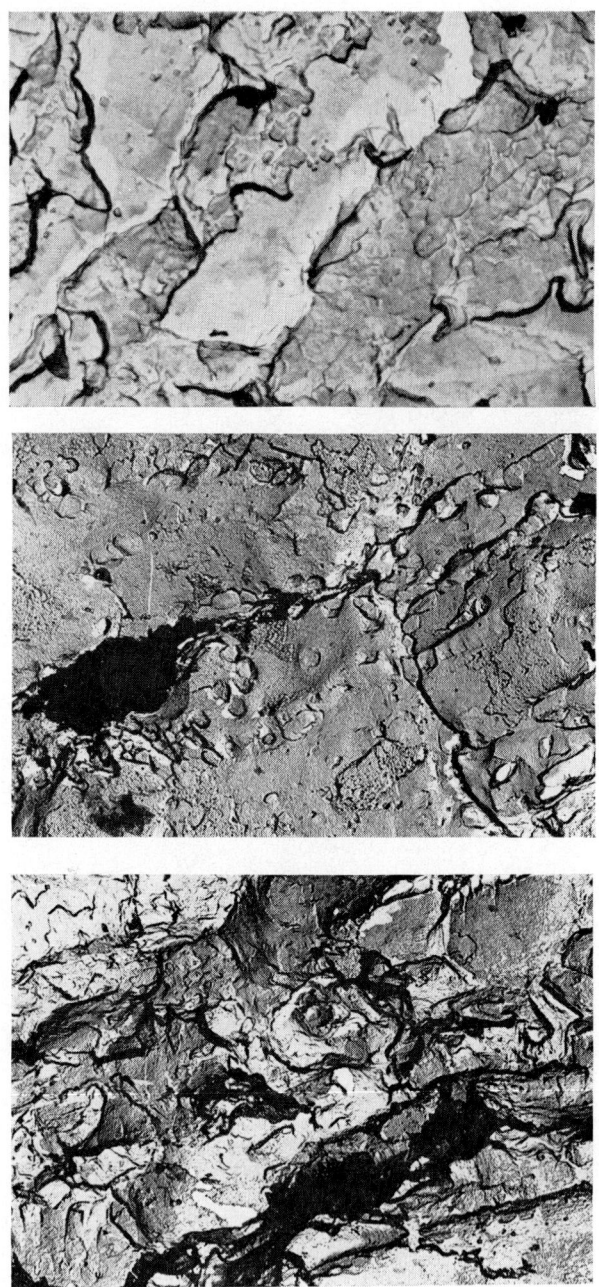

(*top*) ×10,500.
(*center*) ×7500.
(*bottom*) ×10,500.

Etched by immersion in 30 per cent HCl, 20 per cent HNO_3, and 50 per cent H_2O solutions. The direction of crack propagation is parallel with the streaks.

FIG. 10—Electron Micrographs Illustrating the Etched Fracture Face in the Slow Crack-Propagation Region.

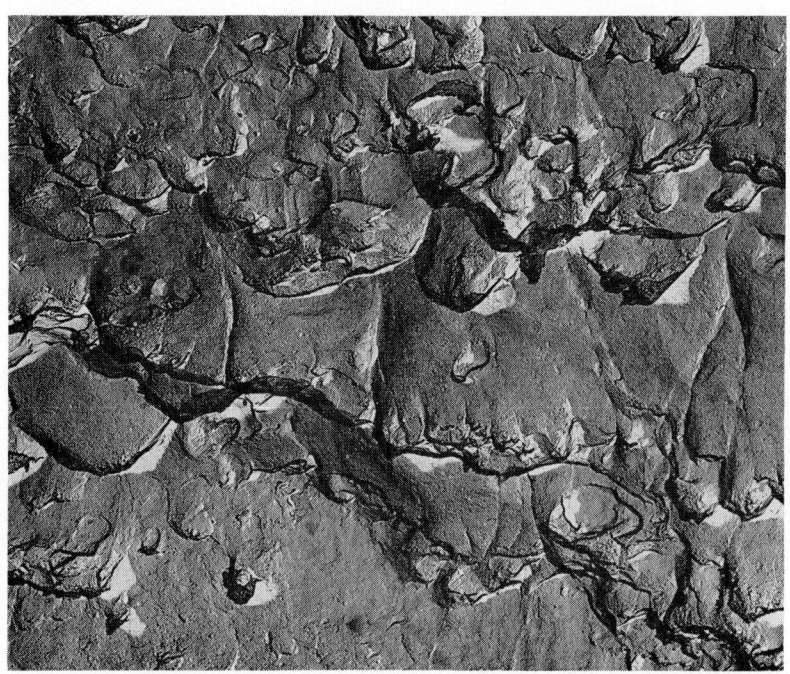

FIG. 11—Electron Micrograph Illustrating Etched Fracture Surface in the Shear Lip Region (final magnification ×16,500).

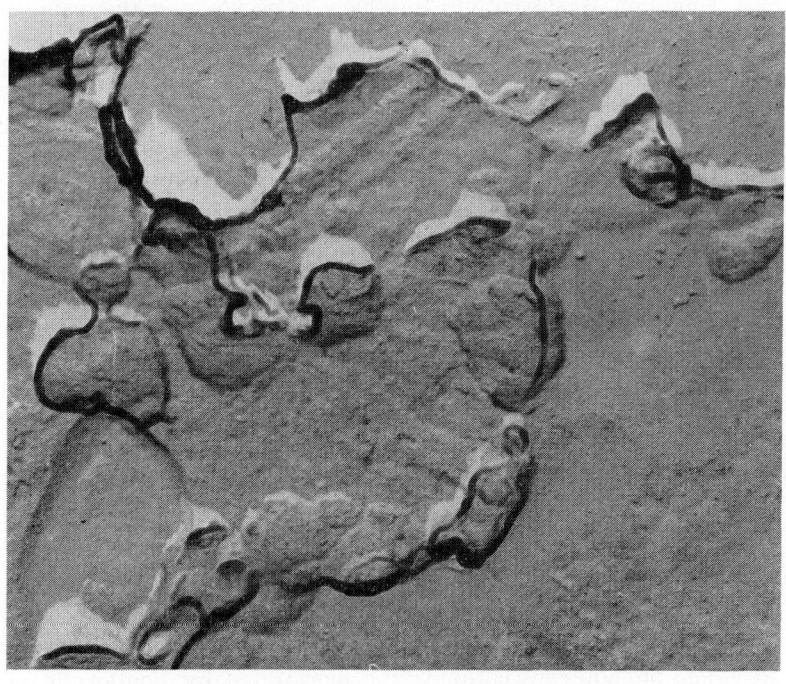

Etched with HCl-HNO_3-H_2O solution.
FIG. 12—Electron Micrograph Illustrating the Profile of the Fracture Surface in the Slow Crack-Propagation Area (final magnification ×16,500).

Figure 11 illustrates two-stage replicas taken from the lightly etched fracture surface of the specimen from the shear lip region. It is interesting to note that the dimples are heavily etched near the apex zones and some precipitate or inclusions are left in relief.

Figure 12 shows 2, two-stage replicas taken from the profile of the fracture surface in the slow crack propagation area. The specimen, which was electroless nickel-plated, was mounted and polished over the fracture face. This permitted examination of the complete fracture surface, following the path of crack propagation. It will be noted in the electron microstructure that the crack propagates along the boundaries where chromium carbide precipitates.

Discussion and Conclusions

The observation presented above clearly indicates that AM 350 steel in an SCT condition is susceptible to stress corrosion cracking in natural atmosphere in the proximity of the oceans. The microfissures appear to originate from surface pits and penetrate into the material primarily along delta ferrite or prior austenite boundaries. Delta ferrite phase usually forms stringers along the rolling direction and has a preferred orientation. In transverse specimens, groups of microfissures form in positions approximately parallel with the main fracture and perpendicular to the direction of tensile load. These are joined by the breaking down of the bridges of metal separating them. Discontinuities, or streaks, are left on the fracture face.

The precipitation of chromium carbide and depletion of chromium content, primarily in the delta ferrite boundaries during short-time anneal or conditioning treatment, apparently causes sensitization of steels and promotes susceptibility to corrosive attack (7). Stress-corrosion cracking originates by the formation of microfissures along the delta ferrite or prior austenite boundaries as a result of electrochemical (or electrochemical and mechanical) action. When the microfissures sufficiently penetrate into the metal they are joined together, with a change in the general direction of the fissure, by the breaking down of the bridges of metal separating them.

The joining of microfissures near the fracture origin in very slowly crack-propagating regions occurs by cleavage and decohesion along glide planes of the matrix crystals or along the austenite grain boundaries bridging the microfissures. The steps shown in Fig. 8 indicate the initial stage of the cleavage process. It is not clear whether or not the cleavage and decohesion of crystals result from the embrittlement of the matrix material between the microfissures. If the applied stress is relatively high, as occurs after the main crack has propagated to a certain length in spring-loaded specimens, the breaking down of the bridges of metal separating the microfissures becomes more and more by mechanical shear. This point has clearly been demonstrated by analysis of dimple formation in the fracture faces.

Crussard et al (8,9,10) have shown that, in precipitation hardening and tempered martensitic steels, elongated dimples generally occur by the formation of internal voids at the interface between the matrix and precipitates or inclusions. The voids grow and are progressively distorted by local shear. Each void appears as a dimple on the fracture face with its axis of symmetry parallel to the shear force. They are connected together producing a local shear surface between the microfissures.

The crack propagation most likely occurs in a series of short bursts, as demonstrated by Fontana et al (11) in stress-corrosion cracking of austenitic stainless steels in chloride waters. No fine structural details could be detected in the electron micrographs to identify the arresting points of microfissures and of the main crack front in the brittle fracture zone.

The development of fibrous structure in the ductile regions also appears to occur by formation of microcracks along delta ferrite boundaries and joining by local shear of the bridges of metal separating them (Fig. 7). Apparently depletion of chromium content and the presence of carbide particles in the delta ferrite boundaries somewhat lowers the mechanical strength, promoting an easy path for crack propagation.

Acknowledgments:

Information contained herein is based on a test program conducted at Astropower, Inc. for Douglas Aircraft Co., Inc., under ASD Contract AF33(657)-8543. "Screening Test Program for Evaluation of the Stress Corrosion Susceptibility of Alloys Under Consideration for Application as Skin Material." McDonnell Aircraft Corporation also participated in the above contract as a major subcontractor.

This contract is part of a national undertaking sponsored by the Federal Aviation Agency, with administration and support provided by the Department of Defense, Aeronautical Systems Division, Air Force Systems Command. Basic research and technical support are provided by the National Aeronautics and Space Administration.

Screening tests for evaluation of the stress-corrosion susceptibility of alloys under consideration were directed by Messrs. C. H. Avery and R. V. Turley of Douglas. Thanks are due to Mr. R. Ingersoll for assistance in electron microscopy.

References

(1) R. H. Raring, V. R. Voorhees, Richard H. Raring, J. W. Freemen, and J. W. Schultz, "Progress Report of the NASA Special Committee on Materials Research for Supersonic Transports," *NASA TND-1798*, May, 1963.
(2) G. C. Deutsch, "Materials for a Supersonic Transport," Journal, *Metals*, Vol. 15, 1963, p. 185.
(3) P. H. Denke, "Materials for the Supersonic Transport," Golden Gate Metals Conference, San Francisco, February, 1962.
(4) J. G. Hines and R. W. Hugill, "Metallographic and Crystallographic Examination of Stress Corrosion Cracks in Austenitic Cr-Ni Steels," *Metallurgy of Stress Corrosion Fracture*, T. N. Rhodin, Editor, Interscience Publishers, Vol. 4, 1959, p. 193.
(5) N. A. Nielsen, "The Role of Corrosion Products in Crack Propagation in Austenitic Stainless Steel," Electron Microscope Studies, Interscience Publishers, Vol. 4, 1959, p. 143.
(6) E. R. Parker, "Theory of Brittle Fracture and Criteria for Behavior at Low Temperatures," *Effect of Temperature on the Brittle Behavior of Metals*, ASTM STP 158, Am. Soc. Testing Mats., 1953, p. 116.
(7) "Wrought Stainless Steels," American Society for Metals, Committee on Wrought Stainless Steels, *Metals Handbook*, Vol. 1, 1961, p. 408.
(8) C. Crussard, "A Comparison Between Ductile and Fatigue Fractures," *Proceedings*, International Conference on Fracture, John Wiley & Sons, New York, N. Y., 1959, p. 524.
(9) N. A. Tiner, "Fractographic Analysis of AISI 3440 Steel by Optical and Electron Microscopy," *Proceedings*, Am. Soc. Testings Mats., Vol. 61, p. 800.
(10) D. C. Ludwigson and A. M. Hall, "The Physical Metallurgy of Precipitation-hardenable Stainless Steels," Defense Metals Information Center, *DMIC Report III*, April, 1959.
(11) W. W. Kirk, F. H. Beck, and M. G. Fontana, "Stress Cracking of Austenitic Stainless Steels in High Temperature Chloride Waters," *Physical Metallurgy of Stress Corrosion Fracture*, T. N. Rhodin, Editor, Interscience Publishers, Vol. 4, 1959, p. 227.

APPLICATION OF A CONDUCTING MASK FOR THINNING METALLIC FOILS FOR ELECTRON TRANSMISSION MICROSCOPY

By Thomas A. Despres[1]

Synopsis

A technique is described by which thin film specimens can be prepared from a 0.002 to 0.006-in.-thick piece of metal very rapidly and with considerable improvement in handling facility. This technique has been successfully applied to aluminum, stainless steel, copper, Inconel X, nickel, and nickel-cobalt giving relatively large uniformly thinned areas. Application of a metallic mask and its effect on uniformity of electropolishing is discussed. Dependence of the degree of uniformity obtained in the thin foils during electropolishing upon (1) frontal area, and (2) thickness to frontal area ratio is suggested.

Considerable effort has gone into development of techniques for thinning metallurgical specimens for examination by transmission electron microscopy. In 1949 Heidenreich (1)[2] first showed that it was feasible to obtain transmissible films by electropolishing 0.005-in.-thick aluminum foil. His technique consisted of polishing aluminum and aluminum-copper specimens 0.125 in. in diameter masked by a nonconducting disk.

Since Heidenreich's work, many techniques employing methods of sacrificial thinning have been used wherein relatively large areas of the initial material are dissolved by concentrated preferential attack, leaving small regions from which the thin specimens are cut for examination. This preferential attack is gained through electrode shape and location (2), nonconducting lacquers, type of electrolyte, or combinations of the above.

Fisher and Szirmae (3) have modified the Bollman technique (2) by applying a conducting mask to relatively large, 0.5-in.-diameter specimen. This method, termed the uniform field method (4), also used by Presland (4) has the advantage of minimizing edge current density concentrations which are present in the methods utilizing a nonconducting mask. In this technique localized current density concentrations occur, resulting in perforations. These perforations can either be allowed to grow until two or more of them nearly meet, leaving a thin region from which the specimen can be taken, or, if the polishing is stopped in time, can be cut for examination from around the perforations.

It has been purely a matter of experience as to which of the above techniques proves most successful for an individual experimenter. The one problem common to all methods of thinning is control. The primary dependent variable that must be controlled is the local rate of thinning. It seems reasonable that the control of the complex phenomena affecting this rate would depend

---
[1] Assistant professor of mechanical engineering, Engineering Division, University of Michigan, Dearborn Campus, Dearborn, Michigan.
[2] The boldface numbers in parentheses refer to the list of references appended to this paper.

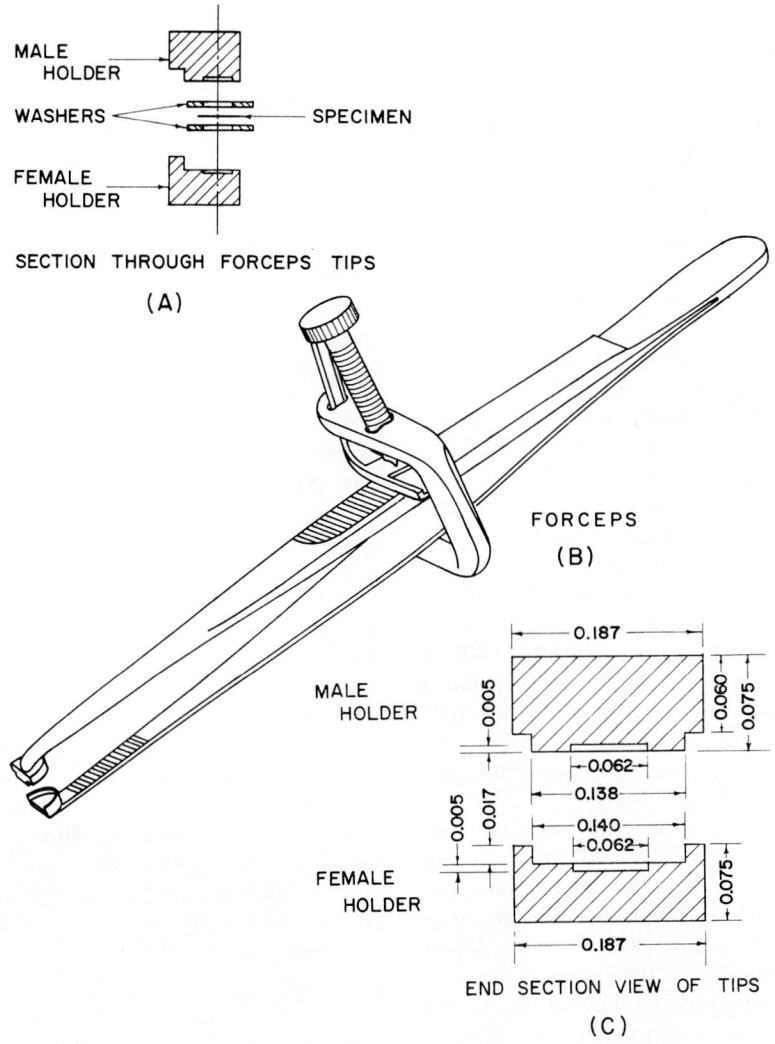

Fig. 1—Sketch of Special Forceps Used for Sandwiching Operation.

not only on electrolyte, voltage, cathode distance and gross average current density but to a greater degree upon irregularities present on a particular area. One means of controlling this independent variable is to minimize the working area.

A very important, purely mechanical problem inherent in most methods, except that of Strutt (5), which requires a special specimen holder, is the handling of the specimen once it is thinned. Washing, cutting, and placing in the microscope holder must be performed with care lest the handling itself affect the structure of the metal.

It would thus be desirable to develop a method of thin film preparation that would (1) produce thin films rapidly, (2) provide ease in handling and storing, and (3) provide for statistical representation of structure.

The author is presently using a method that offers promise toward attaining these desired advantages. This method incorporates handling facility with uniform field considerations to give large regions of thinned area. In addition, the time required for thinning is substantially reduced.

Technique

The technique essentially consists of electropolishing a relatively thin (less than 0.001-in.-thick) specimen between two thin (0.003 to 0.005-in) washers with an I.D. = 0.060 in. and an O.D. = 0.138 in. to the thickness range required for electron transmission studies.

Assume, as is usually the case, that the metal for study is initially 0.002 to 0.006-in.-thick. This piece could either have been rolled to a 0.002 to 0.006-in. thickness before mechanical, thermal, or radiation treatment, or machined from a larger piece by chemical (6) or electrolytic machining (5,6) operations after this treatment.

The specimen material is brought to a thickness of less than 0.001 in. by electropolishing. The methods of thinning previously mentioned or those discussed by Thomas (7) or Kay (8) are easily adapted to this step. When the specimen material is approximately 0.00075-in.-thick, 0.1-in.-square specimens are cut from it with a steel scalpel. These specimens are then sandwiched between the two washers, as shown in Fig. 1 (*top, left*), with the aid of the special forceps (*center*). It is necessary that the washers be clean and free of burred edges. The washers are then spotwelded together near their periphery at two diametrically opposite points on a small capacity bank spotwelder while being held in the forceps. Other means of bonding the sandwich may be employed.

The specimen is next electropolished to the final desired thickness while holding the sandwich in forceps which act as the anode lead. This pair of forceps is insulated, except for the tips where electrical contact with the sandwich is maintained. The electrolyte mixture is held in a small glass container in which a cathode is placed. The geometry of the cathode appears to be of little consequence. The specimen is held in the electrolyte for 10 to 15 sec, removed, washed, and then held over an intense light source to estimate its thickness. The process is continued until the specimen is sufficiently thin for transmission examination. The complete sandwich is then placed in the microscope.

Selection of voltage in the thinning operation is dependent upon the material and the electrolyte. Higher voltages require careful attention in order to determine when to arrest the polishing. The total time required for the final thinning operation is less than 1 min for most materials.

The washers used in the development of this method were made from 304 stainless steel. This material has favorable magnetic and welding characteristics. The washers were made by a simple combination punch and die and electropolished to remove burrs from the inside circumference.

Handling and Preparation Facility

Incorporation of the support frame in this technique appears to eliminate most of the handling problems. The final thinning, washing, and handling operations are performed in the protection of a reinforcement frame, and insertion into the electron microscope holder can be made simply by dropping the complete sandwich into the specimen holder cap, requiring no modification of the microscope specimen holders.

The only major handling problem occurs during the sandwiching operation, but with use of the special forceps shown in Fig. 1 and moderate caution, a specimen can be sandwiched and welded in about 1 min. Extreme care must be taken not to distort the specimen material during the sandwiching operation.

When the small squares are cut from the specimen material, there will be slight distortion present near the cut edges. These edges will be relatively far away from the region to be thinned for examination and should have no effect on it.

The possibility that the washers and welding operation can induce structure changes in the specimen is minimized by the precautions taken in preparation of the washers and in design of the forceps used in the sandwiching operation. Here it must be emphasized that, if bonding is to be done by spot welding, the minimum required heat input should be used.

The use of the "reinforcement frame" and the rapidity of final thinning lead to advantages in handling and storing of specimens. It becomes possible to make up a number of sandwiched specimens, store them in a dessicator, and use them at the microscopist's convenience. Because of the thickness of the washer, the sandwich can be placed on a flat surface without damaging the region to be examined. The specimen can be finally thinned, washed, dried, and inserted in the microscope very quickly. This facility in handling can allow for use of the electron microscope itself as a means to establish whether or not a specimen is thin enough, since the specimen can quickly be removed, further electropolished, and reinserted in the microscope.

Uniformity of Thinning and Current Density Considerations

During the final thinning operation, it appears, at least qualitatively, that the degree of uniformity obtained in thinning is dependent upon (1) the amount of frontal area, and (2) the ratio of initial thickness to frontal area of the film. The frontal-area dependence is supported by the relatively slow growth of perforations which occur during the polishing process. These perforations are regions of high current density which result from irregularities in the material surface or in the anolyte layer caused by the electropolishing action. Working with smaller areas apparently reduces the effect that an irregularity has on uniformity of the field. In addition the anolyte layer should be more uniform with decreasing frontal area of the film. The thickness-area ratio dependence is explained by the fact that the longer time required for thinning a thicker specimen allows for greater deviations from a flat interface. If, for example, one seeks a large area with thicknesses in the range of 0.0005 to 0.0006 in., it can be achieved much more easily if the initial material is 0.001 in. than if it is 0.005 in. In order to gain the same final area with the thicker material and identical conditions, a larger initial area is required, since the deviation from the 0.0005 to 0.0006-in. range requirement would be greater. This reasoning, used to explain results obtained with specimens initially 0.001 to 0.004-in.-thick, leads one to seek a method that would in some way allow for a small frontal area to satisfy condition (1) and a small initial thickness which will satisfy condition (2), hence the requirement of an intermediate polishing operation to bring the specimen material from the range of 0.002 to 0.006-in. thickness down to a thickness approximating 0.00075 in.

Because a nonconducting mask causes preferential thinning at the mask-specimen boundary, its use on very small specimens is not advantageous. The effect of a metallic mask in an arbitrary field, on the other hand, has the ad-

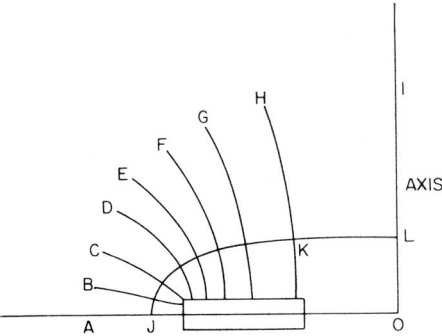

Fig. 2—Two-Dimensional Graphical Field Map Showing the Approximate Current Density Pattern in a Symmetrical Field.

reasoning. Preliminary investigations tend to support this idea.

For the particular geometry and dimensions of the washers chosen, Professor A. D. Moore[3] has supplied approximate maps of the current density patterns which are reproduced in Figs. 2 and 3.

Professor Moore[4] lends further support to the above reasoning with a statement based on these maps:

The effect of the washers is to shape the electropolishing current field so as to insure quite uniform current density over a large

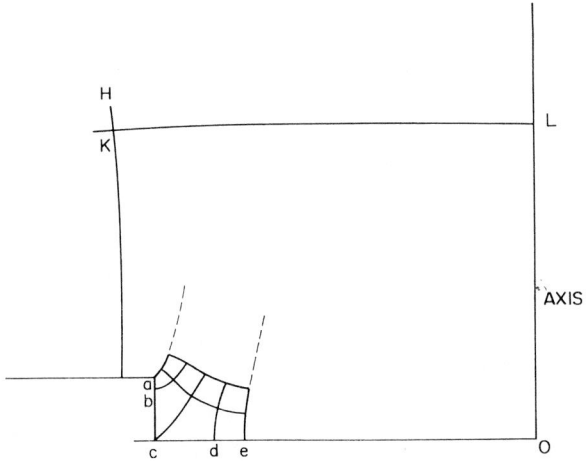

Fig. 3—Detail Map of Field Near Washer Showing Paths of Equal Current Density.

vantage of gaining a current density pattern more favorable to uniform thinning. From investigation of the approximate current density pattern in a symmetrical field, as shown in Figs. 2 and 3, it appears as if an optimum inside diameter and thickness variation of the metal washer can be found for a particular metal where the edge effect will be equalized with the center current density to give more uniformity and better control in thinning. At this writing, work is being done to evaluate this

area of the specimen. The current is three-dimensional, and difficult to plot. Figure 2 is a reasonably close graphical field map—not accurate, but amply good enough to show the facts. If the lines $A, B \cdots G, H$, are revolved about the axis, they mark out tubes of equal current. JKL is an equipotential surface. The KL part of it is so nearly flat, and parallel to the specimen, that specimen surface current density would be almost constant, were it not for distortion caused

[3] Department of Electrical Engineering, University of Michigan.
[4] Personal communication, University of Michigan.

by the presence of the shoulder on the washer.

In Fig. 3, a detail of the map has been drawn near the washer. Here again, the current tubes ab, bc, cd, de, are tubes of equal current. It is seen that on the specimen, the tube end cd is wider than de, and has a much lower current density. Moreover, the density approaches zero as the corner at c is approached. Out to about e, the washer acts as a guard ring, preventing undue thinning of the specimen near the washer.

For all of the area of radius oe, current density will not vary from average value by more than ± 5 per cent.

It is also seen from field considerations that the holding position of the anodically charged forceps will have an affect on the current density pattern and the uniformity of thinning. Better results have been obtained in the final thinning operation by holding the sandwich with the forceps on the outer periphery at the extremes of a diameter.

Results

Since the technique in its present state of development has not yet been used extensively to study mechanisms, dislocation arrays and other structural phenomena, results are primarily based on visual observation in the electron microscope. Several metals, namely, aluminum, stainless steel, Inconel X, copper nickel, and nickel-cobalt were thinned and examined. In all of these metals large thin areas were easily produced. These thin areas had thickness variations which allowed for varying degrees of detail study. In some cases the total area of the specimen across a diameter of 0.060 in. was transmissible to electrons. Although this area consisted of thicker and thinner regions, most of it was uniform and sufficiently thin to allow high magnification studies.

Adaptation to Other Materials and Future Development

It appears that this technique can be adapted to other materials. Development for a particular material should be rapid, since the time required for final thinning is short enough to allow for rapid evaluation of the variables involved. Selection of optimum voltage, electrolyte, and washer diameters will be necessary for any material. For some materials it may be necessary to provide added support by using large mesh grids in conjunction with the washers.

Although cathode size, shape, and location do not appear to have much effect, these may, with development of more efficient methods for observation during the final thinning operation, allow for a means of gaining better uniformity through concentrated attack on the thicker regions.

Acknowledgments:

The author wishes to express appreciation to Prof. W. C. Bigelow for his time, encouragement, and discussions on the present state of the art, to Prof. A. D. Moore for the field maps and statement based on them, and to the Mechanical Engineering Department, University of Michigan, Ann Arbor Campus.

References

(1) R. D. Heidenreich, "Electron Microscope and Diffraction Study of Metal Crystal Textures by Means of Thin Sections," *Journal of Applied Physics*, Vol. 20, 1949, p. 993.

(2) W. Bollman, "Interference Effects in Electron Microscopy of Thin Crystals," *Physical Review*, Vol. 103, 1956, p. 1588.

(3) R. M. Fisher and A. Szirmae, "Observations of Dislocations in Thin Foils of Stain-

less Steel with the Electron Microscope," *Electron Metallography, ASTM STP 262*, Am. Soc. Testing Mats., 1959, p. 103.
(4) D. Kay, *Techniques for Electron Microscopy*, Thomas, Illinois, 1961, p. 241.
(5) P. R. Strutt, "Preparation of Thin Metal Foils from Ordinary Tensile Specimens for Use in Transmission Electron Microscopy," *Review of Scientific Instruments*, Vol. 32, 1961, p. 411.
(6) *Product Eng.*, Nov. 12, 1962, p. 69.
(7) G. Thomas, *Transmission Electron Microscopy of Metals*, Wiley, N.Y., 1962, pp. 150–159.
(8) D. Kay, *Techniques for Electron Microscopy*, Thomas, Illinois, 1961, pp. 235–245.

INVESTIGATION OF MICROSTRUCTURE AND ROOM-TEMPERATURE FRACTURE IN SM-200 NICKEL-BASE ALLOY

By R. W. Hertzberg[1] and J. A. Ford,[2] Associate Member ASTM

Synopsis

The deformation and fracture behavior of SM-200, a nickel-base alloy, was studied at room temperature under tension and flexural loading. Electron fractographic techniques were extensively used to determine which phases and fracture mechanisms contributed to the fracture process. In addition, an investigation of the phase constitution of SM-200 using electrolytic phase extraction was conducted in order to relate mechanical behavior to alloy microstructure. A solution heat treatment was employed in an attempt to produce a microstructure exhibiting properties superior to those shown by as-cast material.

Failure was found to occur along subgrain boundaries containing titanium carbide particles and massive γ' clusters in as-cast materials, while TiC and $M_{23}C_6$ particles were considered responsible for subgrain boundary fracture in solution heat-treated specimens.

In recent years, the need for higher operating temperatures during turbine engine performance has led to the development of highly complex nickel-base alloys. To optimize strength level and rupture life at elevated temperature, many alloying elements have been introduced to take advantage of all known useful strengthening mechanisms (solid solution, precipitation, and dispersion hardening). As a result, it has become extremely difficult to clearly define the operative fracture mechanisms in these multi-component, multi-phase alloys.

One problem found common to many such alloys has been that of low room-temperature ductility. Accordingly, this investigation was undertaken to gain an understanding of the relation between room-temperature mechanical behavior and microstructure of SM-200 nickel-base alloy.

Experimental Procedure

Phase Identification:

The characteristics of the secondary phases present in SM-200 (Table 1) were studied using optical and electron microscopy and X-ray and electron diffraction. Specimens were mechanically polished through 0.5μ diamond, vibratory polished with Linde B alumina and swab etched in either Kalling's reagent or the University of Michigan etch copper chloride (50 g ($CuCl_3$), 360 ml acetic acid, 230 ml concentrated hydrochloric acid (HCl), 50 ml concentrated sulfuric acid (H_2SO_4), and 1 g chromic acid). The secondary phases were selectively removed from the matrix using four complementary

---

[1] Formerly associate research scientist, advanced metallurgy, United Aircraft Corp., Research Laboratories, East Hartford, Conn. Presently, research assistant, Department of Metallurgy, Lehigh University, Bethlehem, Pa.

[2] Senior research scientist, Advanced Metallurgy, United Aircraft Corporation, Research Laboratories, East Hartford, Conn.

techniques; electrolytic extraction in 10 per cent aqueous or alcoholic HCl and 20 per cent aqueous phosphoric acid (H_3PO_4), chemical etching in 10 per cent alcoholic bromine, and extended etching in the University of Michigan etch. The extracts were centrifuged and washed in absolute methyl alcohol and vacuum dried at room temperature. X-ray diffraction studies, using a 114.59-mm diameter powder camera, and fluorescent analyses were made of extracts immediately after drying. Selected-area electron diffraction of the small grain boundary precipitates produced by heat treatment was performed on extraction replicas.

TABLE 1—NOMINAL COMPOSITION OF SM-200 NICKEL-BASE ALLOY.

Element	Per Cent
C	0.12 to 0.17
Mn	0.20 max
S	0.015 max
Si	0.20 max
Cr	8 to 10
Co	9 to 11
W	11.5 to 13.5
Nb	0.75 to 1.25
Ti	1.75 to 2.25
Al	4.75 to 5.25
B	0.01 to 0.02
Zr	0.03 to 0.08
Fe	1.5 max
Cu	0.10 max
Ni	balance

Heat Treatment:

Solution heat-treating experiments were carried out in evacuated quartz ampules at temperatures up to 2400 F and times up to 24 hr. The ampules were water quenched at the end of each heat treatment; however, the ampules were not broken by the quench. As-cast and solution heat-treated specimens for mechanical testing were given a simulated coating treatment of 4 hr at 2000 F in argon followed by an air cool.

Replication Techniques:

Replicas of fracture surfaces were prepared using Faxfilm (cellulose acetate) and carbon. A strip of Faxfilm moistened with acetone was pressed onto the fracture surface and permitted to dry. After being stripped off, the Faxfilm tape was placed in a vacuum chamber and shadowed with platinum at an angle of 45 deg. A layer of carbon was then deposited while the tape rotated on a table inside the chamber. The layer of Faxfilm was dissolved by placing the compound replica into a bath of acetone.[3] After changing the bath two or three times and

FIG. 1—As-Cast Microstructure of SM-200 (× 100).

allowing the carbon replica to float freely for 10 to 20 min, the replica was picked up on a 200-mesh grid. This procedure yielded clean replicas, often representing over 90 per cent of the fractured surface. Replicas from polished and etched surfaces were prepared using collodion, backed with carbon and shadowed with platinum. The collodion was subsequently dissolved in amyl acetate.

Mechanical Testing:

Test specimens were subjected to both flexure and tension loading. A three-point bending apparatus was placed on a

---
[3] C. Beachem, private communication.

TABLE 2—SUMMARY OF d VALUES FOR PHASES FROM SM-200 IN THE AS-CAST AND CREEP-TESTED STATES.

Ni₃Al[a] d(I/I₁)	TiC[a] d(I/I₁)	Cr₂₃C₆[a] d(I/I₁)	62-133-E7 10% Alcohol HCl d(I)	As-Cast 62-133-E3 20 per cent H₃PO₄ d(I)	62-133-E12 University of Michigan Etch d(I)	Creep Tested 700 hr at 1700 F/23.5 ksi 62-523-E1 10% Alcoholic Bromine d(I)
3.60 (40)	...	...	...	...	...	...
...	...	3.21 (25)	...	...	...	...
...	...	3.07 (25)	...	...	...	...
...	...	2.66 (50)	...	...	...	2.90 (W)
2.54 (40)	...	...	...	...	...	...
...	2.508 (80)	...	2.513 (VS)[b]	...	2.504 (S)	2.513 (VS)
...	...	2.44 (25)	...	...	...	...
...	...	2.38 (100)	...	...	...	2.394 (M)
...	2.179 (100)	2.17 (100)	2.175 (VS)	...	2.172 (S)	2.183 (S)
2.074 (100)	...	...	...	...	...	...
...	...	2.05 (100)	2.062 (S)	2.053 (VS)	...	2.055 (MW)
...	...	1.88 (75)	...	...	...	1.897 (W)
1.799 (70)	...	1.80 (100)	1.787 (MS)	1.780 (S)	...	1.810 (W)
...	...	1.77 (50)	...	...	...	...
...	...	1.68 (25)	...	...	...	1.683 (VW)
...	...	1.62 (25)	...	...	...	1.605 (W)
1.603 (40)	...	1.60 (50)	...	...	...	...
...	1.535 (50)	...	1.540 (M)	...	1.537 (M)	1.540 (M)
...	...	1.49 (25)	...	...	...	...
...	...	1.48 (25)	...	...	...	1.475 (W)
1.461 (20)	...	...	...	...	...	...
...	...	1.38 (25)	...	...	...	...
...	...	1.34 (25)	...	...	...	...
...	...	1.33 (50)	...	...	...	...
...	1.311 (30)	...	1.313 (MW)	...	1.314 (M)	1.311 (M)
...	...	1.29 (100)	...	...	...	...
1.265 (60)	1.255 (10)	1.25 (100)	1.268 (M)	1.263 (M)	1.246 (W)	1.261 (M)
...	...	1.23 (100)	1.256 (W)	...	...	1.235 (VW)
...	...	1.22 (25)	...	...	...	...
...	...	1.19 (75)	...	...	...	...
...	...	1.17 (75)	...	...	...	...
1.078 (60)	1.086 (5)	...	1.080 (MW)	1.078 (M)	1.091 (VW)	1.092 (W)
1.032 (40)	...	...	1.034 (MW)	1.025 (W)	...	...
...	0.997 (5)	...	0.998 (W)	...	1.002 (VW)	0.997 (VW)
...	0.971 (30)	...	0.974 (MW)	...	0.975 (W)	0.974 (VW)
0.893 (20)	...	...	0.890 (MW)	0.894 (W)	...	...
...	0.884 (30)	...	...	...	0.890 (W)	0.888 (MW)
...	0.833 (30)	...	0.839 (MW)	...	0.839 (W)	0.839 (W)
0.819 (70)	...	...	0.822 (M)	0.821 (M)	...	...
0.798 (70)	...	...	0.801 (M)	0.801 (M)	...	...

[a] Values from ASTM X-ray Powder Data File (*STP 48*).
[b] V = very; S = strong; M = medium; W = weak.

Leitz light microscope stage, thus allowing the deformation process to be viewed on a polished surface while the test was being performed. The specimens (nominally 1⅛ by 1/16 by 1/16 in.) were cut from airfoil sections of cast turbine blades. Tension tests were conducted at a strain rate of 0.03 in./min using a 60,000 lb capacity Tinius Olsen machine on cast tension bars of 0.25 diameter with a 1.0 in. gage length.

RESULTS AND DISCUSSION

Phase Identification:

The as-cast microstructure of SM-200, Fig. 1, is very similar to that of IN 100[4]

---
[4] S. F. Sternasty and E. W. Ross, "IN-100, a Cast Alloy," *Metal Progress*, Vol. 80, No. 6, December, 1961, p. 73.

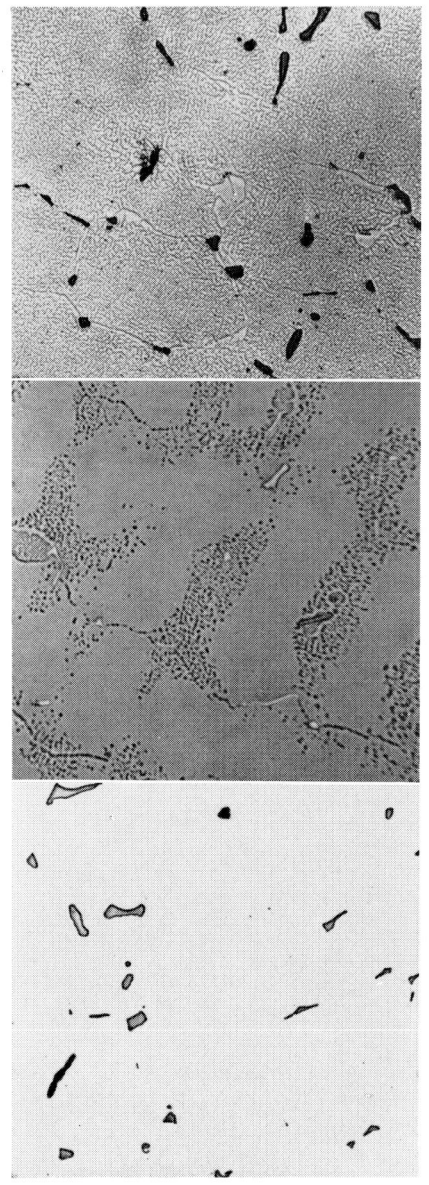

(top) 2000 F
(center) 2100 F
(bottom) 2200 F

Fig. 2—Microstructure After 1-hr Heat Treatment (× 200).

fied as MC carbide, where M denotes various metal atoms, and a massive boundary γ' formed during solidification by eutectic decomposition.[5] The nonuniform etching seen in Fig. 1 indicates that the alloy is segregated and that the microstructure is fine-grained. These small grains are in reality dendrite arms separated by small angle boundaries, as will be discussed later. In general the macrograin size of as-cast SM-200 is large ($\frac{1}{8}$ to $\frac{1}{2}$ in.) unless extreme processing care is taken to produce fine grains.

The extraction of these secondary phases by electrolytic or chemical means provided proof of their nature. Precipitated γ', with a lattice parameter of 3.58 A, was extracted using 20 per cent aqueous H_3PO_4 and was shown by X-ray fluorescent analysis to contain all of the alloying elements detectable by this method which were present in the alloy. The MC carbide phase, extracted using electrolytic HCl, alcoholic bromine, and the University of Michigan etch, was shown by fluorescent analysis to contain titanium, tungsten, and columbium. The results of typical X-ray diffraction analyses of extracts prepared by the four different methods are summarized in Table 2.

Heat Treatment:

The presence of titanium carbide (TiC) and massive γ' in the interdendrite boundaries of cast SM-200 has led to a low-energy fracture through these areas as discussed below. Solution heat-treating experiments were undertaken to alleviate this condition. Figure 2 illustrates the effects of 1 hr at 2000, 2100, and 2200 F. At 2000 F there is little evidence of any γ' solutioning, and it appears that only agglomeration has oc-

and contains the normal precipitated γ', a large hard boundary phase identi-

---

[5] R. A. Gregg and B. J. Piearcey, "Interpretation of the 'Kidney' Constituent in Nickel-Base Superalloys," Pratt and Whitney Report No. *AMRDL 62-005*, Dec. 8, 1962.

curred (*top*). However, at 2100 F (*center*) the γ' within the interior of each dendrite arm has been successfully solutioned, while that adjacent to the dendrite boundaries as well as the massive γ' has not been significantly affected. It is well known that alloying elements, with the exception of iron, raise the stability of γ',[6] and thus the nonuniform γ' solutioning exhibited by SM-200 is strong proof of alloy segregation during solidification. One hour at 2200 F has caused complete solutioning of both the precipitated γ' and the massive γ' (*bottom*).

It can be seen from Fig. 2 that the MC type carbides containing titanium, columbium, and tungsten have not been significantly affected by the solution heat treatments. Measurements of the volume of these carbides, using point-counting techniques, indicated a range of 1.0 to 2.4 volume per cent, which did not depend on the heat treatment. All attempts to put the TiC phase into solution were un-

[6] C. H. Lund, "Physical Metallurgy of Nickel-Base Superalloys," DMIC *Report 153*, Defense Metals Information Center, May 5, 1961.

successful at temperatures through 2400 F. At this temperature there was evidence of local melting in regions in which the massive γ' had existed prior to solutioning.

Precipitation of a grain boundary phase, identified as $M_{23}C_6$ carbide by selected-area electron diffraction of extraction replicas, has occurred at 2000 F. The M atom is most probably chromium plus other similar alloying elements. No evidence of the presence of M_6C carbides

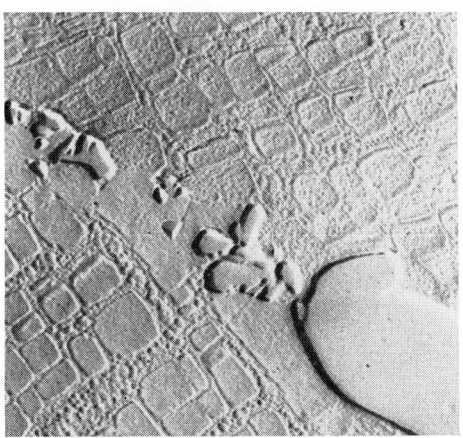

Heat treatment of 3 hr at 2250 F plus 4 hr at 2000 F.
Fig. 3—Grain Boundary Precipitation of $M_{23}C_6$ Within Gamma Prime Film ($\times$ 11,800).

Fig. 4—Brittle Fracture and Decohesion of TiC Particle ($\times$ 500).

has been found at any temperature for times up to 24 hr. In contrast to the isolated discontinuous $M_{23}C_6$ precipitation discussed above, the effect of a 4-hr, 2000 F aging treatment of as-cast plus solution-treated SM-200 is illustrated in Fig. 3. Here it is evident that the combination of solution heat treatment plus aging develops a continuous boundary film of γ' in which $M_{23}C_6$ is discontinuously precipitated.

Mechanical Behavior:

With the aid of the three-point bending apparatus designed to allow simultaneous loading and microscopic examination, it was possible to study the behavior of slip traces and to follow the progression of the fracture as it moved across the polished face of the specimen. From such studies it was revealed that slip traces crossed certain boundaries without any noticeable change in direction or sidewise jog. This would indicate that these boundaries were sub-boundaries or dendrite walls within a larger grain. In most cases these dendrite walls constituted the major portion of the fracture path. Therefore, even though major grain boundaries played a minor role in crack propagation during flexural loading, the general fracture mode for as-cast material may be classified as being intergranular along subgrain boundaries.

It was found that a large percentage of TiC particles were segregated along dendrite walls and major grain boundaries. These particles were seen to crack at low stress levels compared to those required to fracture the specimen. Evidence of the resistance to plastic deformation of this phase was noted by the inability of slip traces to cross any of the particles.

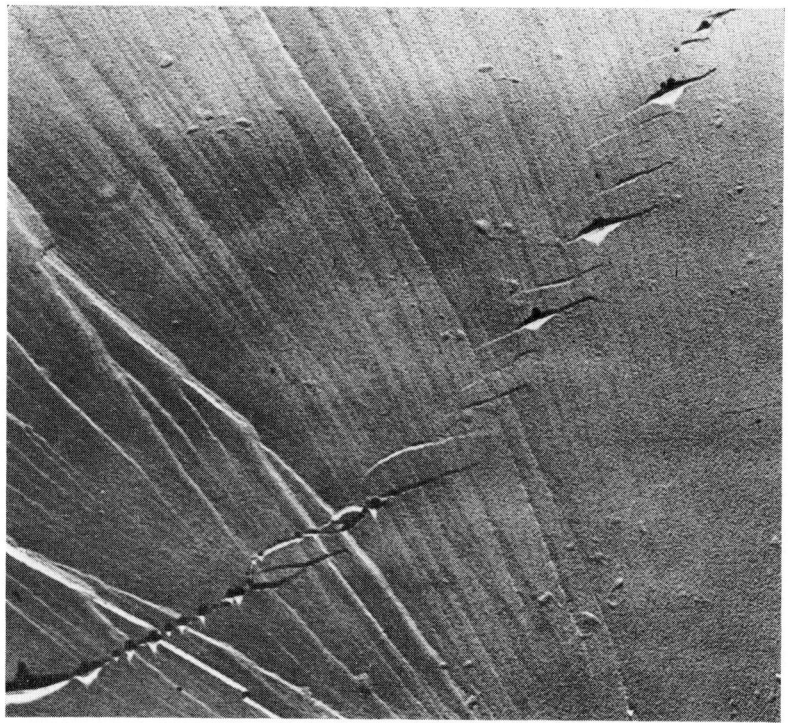

FIG. 5—Electron Fractograph of Cleaved TiC Particle Showing Typical Fan-Like Pattern (× 21,600).

In addition to the brittle behavior of the TiC phase, fracture also occurred at the particle-matrix interface. Figure 4 illustrates both brittle fracture and decohesion of a TiC particle. It was found by replication techniques that failure associated with the TiC phase was a common event on the fracture surface. A striking illustration of a cleaved particle with fan-like cleavage steps radiating from a nucleation site is seen in Fig. 5.

This particle is interpreted as being TiC on the basis of its size and shape. If the fracture process had caused decohesion at the TiC-matrix interface, an internally free surface would be generated. When dislocations approach such a surface, they will produce slip steps similar to those seen on a polished free surface. The fracture produced by particle-matrix decohesion would, therefore, be relatively flat and contain numerous slip markings as shown in Fig. 6.

The other microconstituent considered responsible for low-angle boundary failure of cast SM-200 was the agglomerated form of γ'. Like TiC, the clusters of massive γ' often segregate at dendrite cell walls. During flexure it was noted that the advancing crack either propagated around the periphery or through these clusters. Electron fractographs, Figs. 7 to 9, reveal the presence of these clusters, thus indicating their participation in the fracture process. Note the morphological similarity between Figs. 7 to 9 and the electron micrograph of a cluster shown in Fig. 10. By the smooth surfaces and

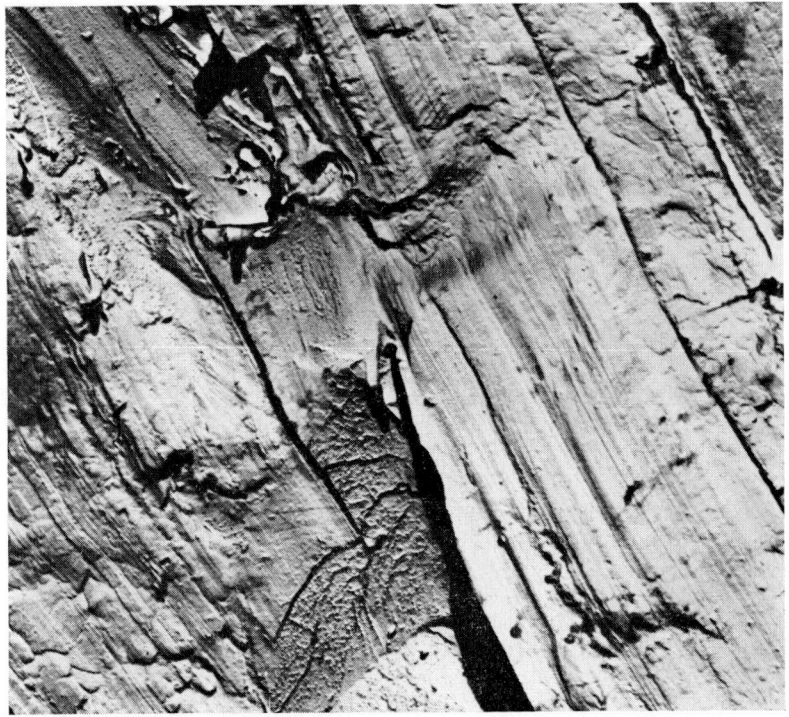

Fig. 6—Electron Fractograph of TiC-Matrix Decohesion Showing Slip Traces on Free Surface ($\times$ 10,800).

general absence of cleavage markings, the fracture mechanism of the clusters shown in Figs. 7 to 9 is interpreted as being that of decohesion, while slip markings indicative of plastic deformation are noted in the nickel solid-solution matrix surrounding the nodules. An indication of the complex fracture behavior of this alloy is shown in Fig. 11, where cleavage, decohesion, and plastic deformation are found in a localized area.

Fig. 7—Electron Fractograph of Decohesion of Massive Gamma Prime-Matrix Interface (× 18,000).

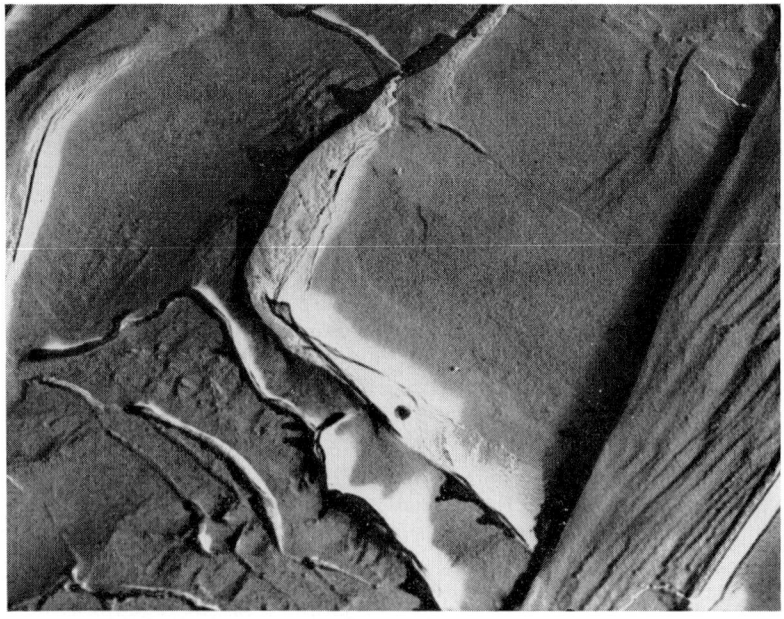

Fig. 8—Electron Fractograph Further Illustrating Decohesion of Gamma Prime-Matrix Interface and Appearance of Slip Traces in Matrix (× 20,000).

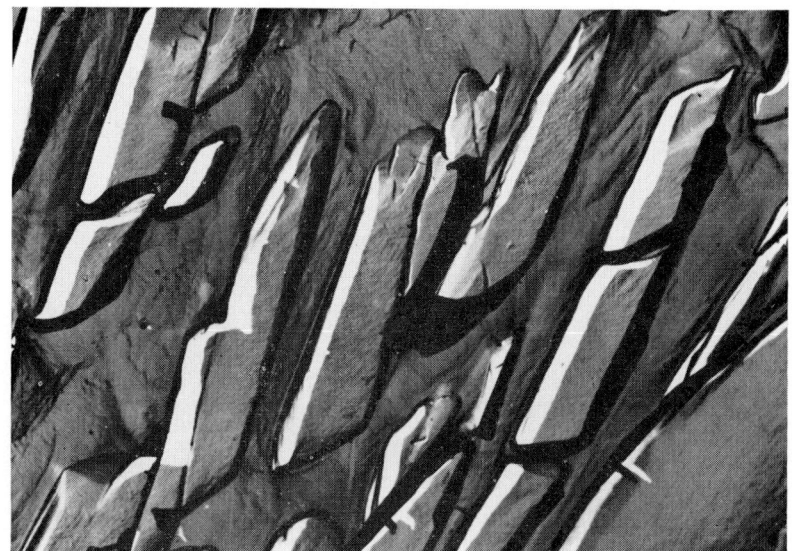

FIG. 9—Electron Fractograph of Decohesion of Massive Gamma Prime-Matrix Interface (× 10,400).

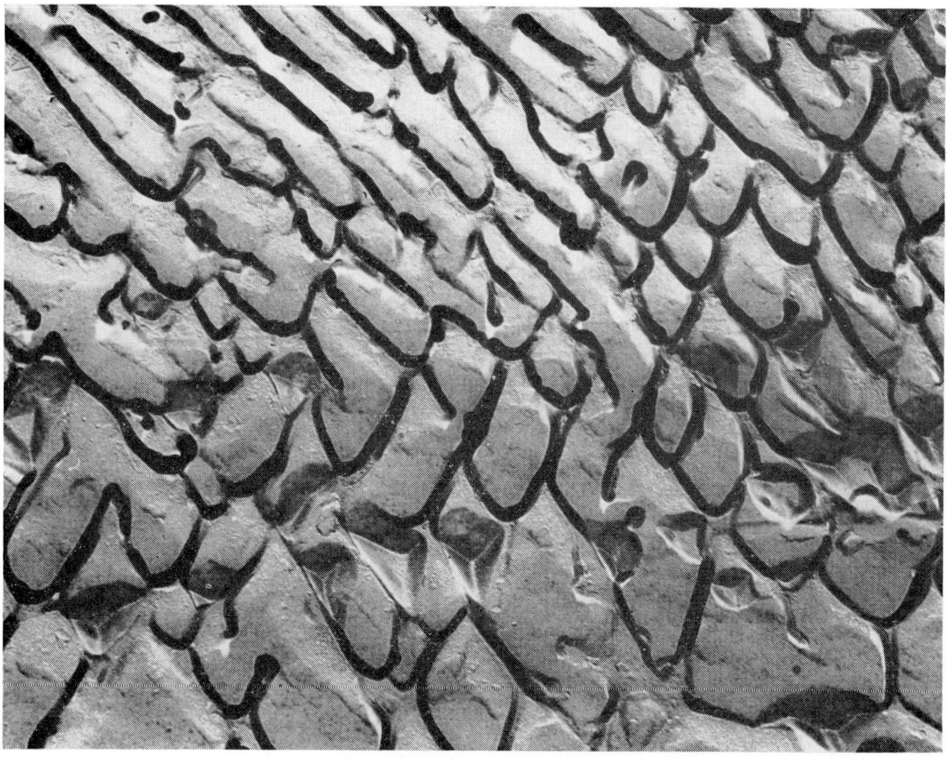

FIG. 10—Electron Micrograph of Massive Gamma Prime Particle Cluster (× 16,000).

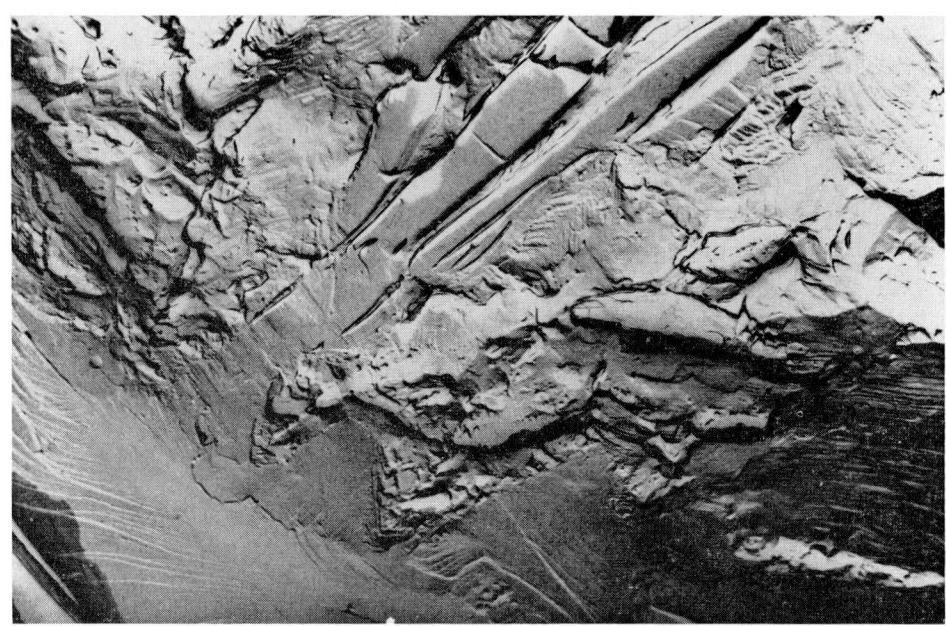

Cleavage, interfacial decohesion, and ductile matrix failure are illustrated.

FIG. 11—Electron Fractograph Showing Complex Nature of Fracture Behavior in SM-200 ($\times$ 10,800). Reduced one third for reproduction.

TABLE 3—TENSILE DATA.

	Aged 4 Hr at 2000 F			Solution Heat Treated 3 Hr at 2250 F and Aged 4 Hr at 2000 F			
Specimen No.	Yield Strength, psi	Tensile Strength, psi	Total Elongation, per cent	Specimen No.	Yield Strength, psi^a	Tensile Strength, psi^b	Total Elongation, per cent
62-580-25	120 700	132 600	3.5	62-580-31	134 000	145 900	4.0
62-580-26	118 800	135 300	4.1	62-580-32	130 500	141 900	2.6
62-580-27	116 900	126 000	2.3	62-580-33	132 900	137 200	1.1
62-580-28	111 900	129 500	4.4	62-580-34	134 600	144 000	3.5
62-580-29	122 200	132 000	2.5	62-580-35	133 800	140 300	1.5
62-580-30	115 700	127 500	3.6	62-580-36	129 200	135 500	2.2
Average	117 700	130 500	3.4		132 500	140 800	2.5

[a] Difference in Yield Strength = 12.6 per cent.
[b] Difference in Tensile Strength = 7.9 per cent.

Solution heat-treating experiments were undertaken in an attempt to eliminate both microconstituents found to be detrimental to room temperature mechanical behavior. All specimens were given the simulated coating cycle of 2000 F (4 hr), while half of them had an initial solution treatment of 2250 F (3 hr). Tension tests and flexural studies were undertaken to study the effect of changes in microstructure on the behavior of the material. The tension test data, summarized in Table 3, indicate an increase in yield and tensile strength

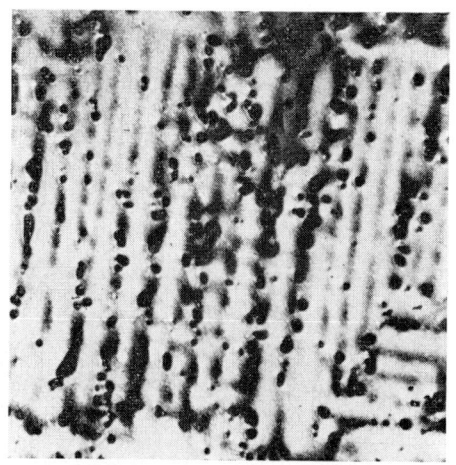

Specimen aged 4 hr at 2000 F.
FIG. 12—Striated Fracture Surface of Tension Bar (× 8).

FIG. 13—Parallel Dendrite Walls Within An As-Cast Grain (× 50).

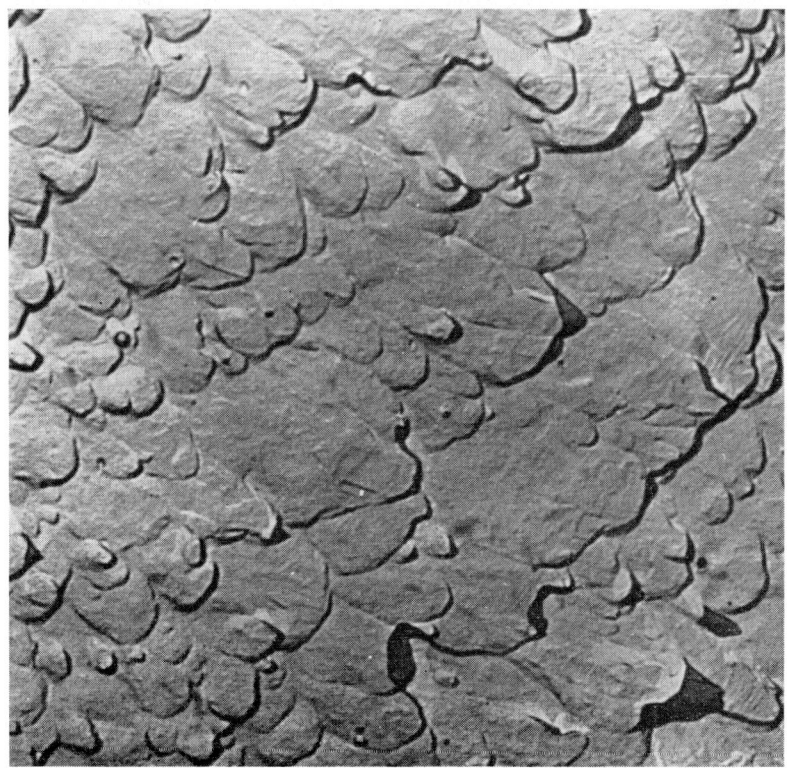

FIG. 14—"Elongated Dimples" Indicating Local Plastic Deformation (× 14,800).

accompanying a reduction in total elongation for the solution-treated and aged specimens. The as-cast and aged specimens also exhibit relatively low ductility. (Though room temperature ductility was apparently decreased by the solution treatment and aging cycle, there are indications that solution-treated specimens possess both increased strength and ductility at elevated temperatures). Creep rupture life at 1400 F increased considerably when the specimens were given the solution anneal.[7] The fracture path for all tension bars was predominantly along dendrite walls. The striations noted on the fracture surfaces in Fig. 12 are interpreted as being caused by failure along parallel dendrite cell walls within each grain. This process can be more readily visualized by referring to Fig. 13, which illustrates the parallelism between neighboring dendrite walls. The microscopic appearance of the fracture surface of

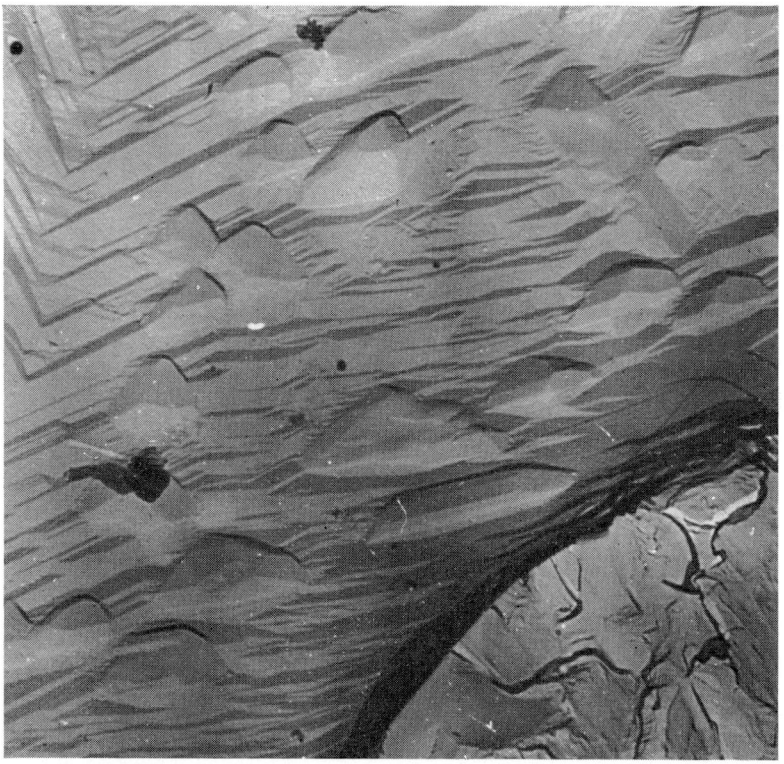

FIG. 15—Decohesion of $M_{23}C_6$ Particles From the Grain Boundary Gamma Prime Film ($\times$ 14,800).

specimens heat treated at 2000 F (4 hrs was similar to that observed for as-cast) material. However, solution-treated and aged specimens did reveal a distinct change in fracture appearance. As expected, no more massive γ' clusters were found, while more indications of local plastic deformation as evidenced by the increasing occurrence of regions consisting of "elongated dimples" were noted

---

[7] A. Pinkowish. private communication.

(Fig. 14). An intergranular fracture mechanism was revealed and associated with the grain boundary film of γ' and imbedded $M_{23}C_6$ carbide particles. Figure 15 illustrates decohesion of the $M_{23}C_6$ particles from the γ' film. The markings associated with the grain boundary γ' are considered to be slip markings rather than cleavage steps in view of the fact that slip lines were seen crossing the grain boundary film during bending.

Conclusions

1. The as-cast microstructure consists of precipitated γ', massive γ' formed during solidification, and TiC within a nickel-base, solid-solution matrix.

2. Nonuniform etching and local variations in response to heat treatment indicate considerable alloy segregation.

3. A 2000 F heat treatment causes precipitation of grain boundary $M_{23}C_6$, while a solution treatment of 2200 F in addition to aging at 2000 F results in the formation of a γ' grain boundary film surrounding the $M_{23}C_6$ particles. Heat treatment at 2200 F also brings about complete solutioning of the precipitated and massive γ' clusters.

4. Room temperature fracture in all specimens occurs predominantly along subgrain boundaries. Metallographic and electron fractographic examinations reveal that massive γ' clusters and TiC particles are responsible for the subgrain boundary failure and consequent low room-temperature ductility in as-cast and aged specimens, while TiC and $M_{23}C_6$ subgrain boundary particles are believed to cause failure in solution-treated and aged test bars.

MICROCONSTITUENTS IN HIGH-SPEED STEEL

By P. K. Koh[1] and H. Nikkel[1]

Synopsis

Electron microprobe analysis can be applied to the study of microconstituents in high-speed steel. *In situ* analyses were made in alloy carbides and manganese sulfide inclusions.

Apparent heterogeneity is shown in the iron and vanadium contents of three neighboring M_6C carbides, and the fact that VC carbides contain considerable amounts of alloying elements other than vanadium is illustrated. In a massive VC carbide no significant composition changes were observed in its center after tempering, and annealing heat treatments were conducted in vacuum.

High-speed steel is one of the best documented and most studied commercial alloys. Phase changes that occur during secondary hardening and tempering of the steel center around the formation of alloy carbides which are generally given the formulas, Fe_4W_2C, Fe_3W_3C, $(Fe, W, Mo, Cr, V)_6C$, or M_6C.

X-ray diffraction analyses of high-speed steel specimens and of residues chemically extracted from them have shown that the crystalline structure of the M_6C carbides is face-centered, cubic with lattice parameters varying from 10.60 to 11.04 A. However, the exact composition of individual carbides cannot be determined by chemical or x-ray methods. Additional alloy carbides and intermetallic compounds, such as $Cr_{23}C_6$, VC and Fe_3W_2, may also be involved in the transformations that occur in some high-speed steels.

It is the purpose of this investigation to apply electron microprobe analysis to the *in situ* study of carbides and other microconstituents in high-speed steel in the annealed, hardened, and tempered states. From the experimental results it may be possible to learn whether the M_6C carbide is homogeneous and stoichiometric, or if it consists of different alloying elements varying over a range of composition. Knowledge of the exact compostion of individual carbides during matrix phase transformation may also be important in the study of red hardness. Another interesting area of investigation is the constitution of other microconstituents, such as sulfides, and how they affect machinability.

Since we started this investigation quite recently, the data collected so far represent only a preliminary survey. Some highlights of general interest follow.

Results

Figures 1 through 5 illustrate heterogeneity in three neighboring M_6C car-

---

[1] Engineers, Physical Metallurgy, Homer Research Laboratories, Bethlehem Steel Co., Bethlehem, Pa.

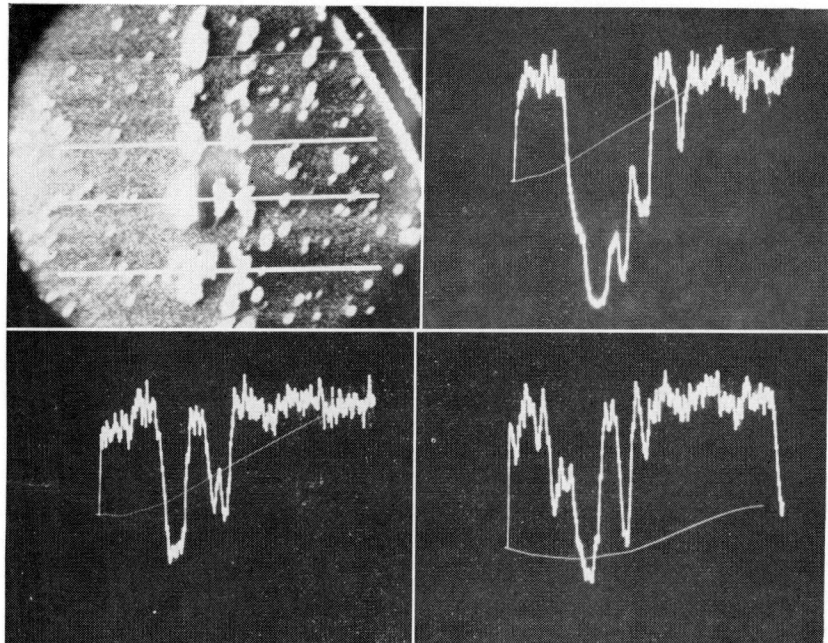

(*bottom, left*) Scan 1   (*top, right*) Scan 2   (*bottom, right*) Scan 3

White lines indicate where three line scans of the microprobe were made. Oscilloscope traces show Fe Kα x-ray intensity along the lines.

FIG. 1—Neighboring Carbides in the Electron Image ($\times$500).

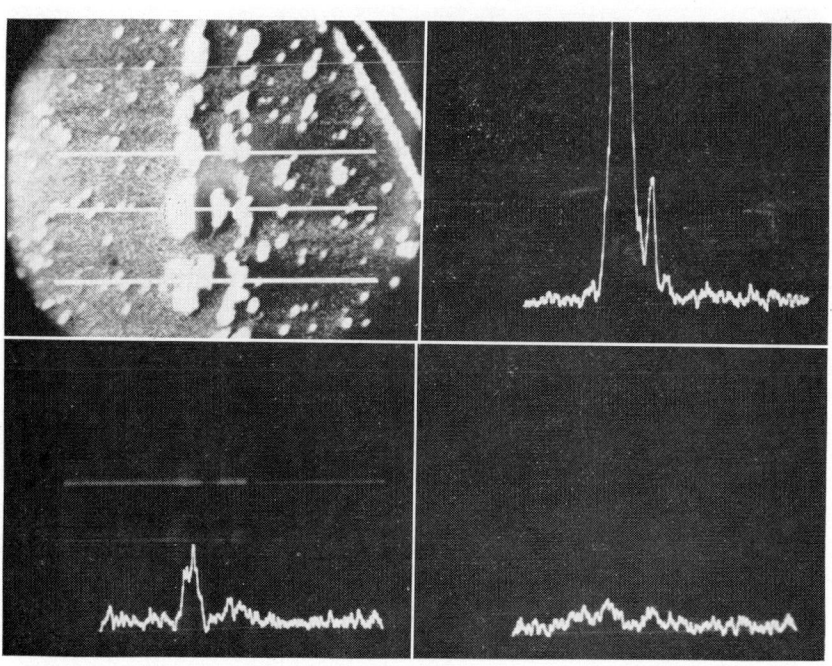

(*bottom, left*) Scan 1   (*top, right*) Scan 2   (*bottom, right*) Scan 3

Oscilloscope traces show V Kα x-ray intensity.

FIG. 2—Neighboring Carbides in the Electron Images ($\times$500).

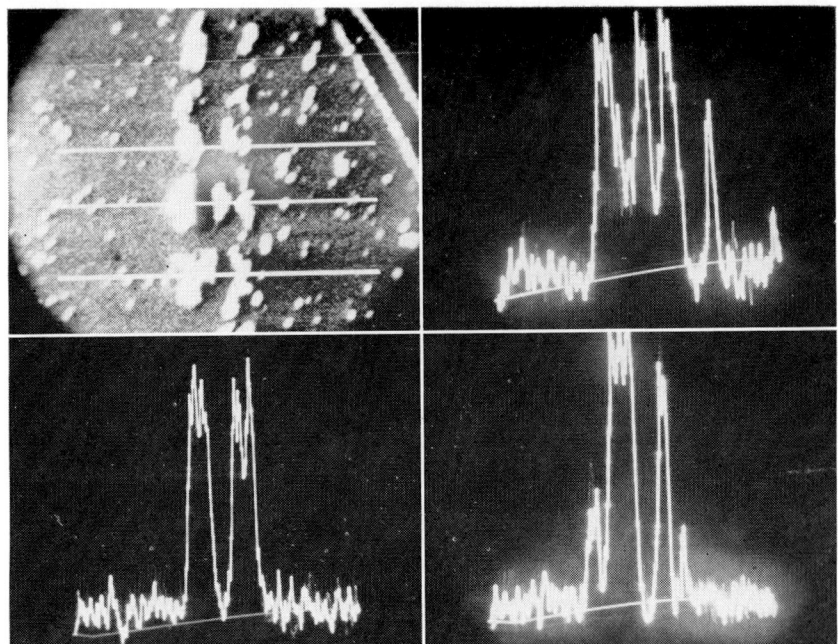

(bottom, left) Scan 1   (top, right) Scan 2   (bottom, right) Scan 3

Oscilloscope traces show Mo Lα x-ray intensity.

FIG. 3—Neighboring Carbides in the Electron Image (×500).

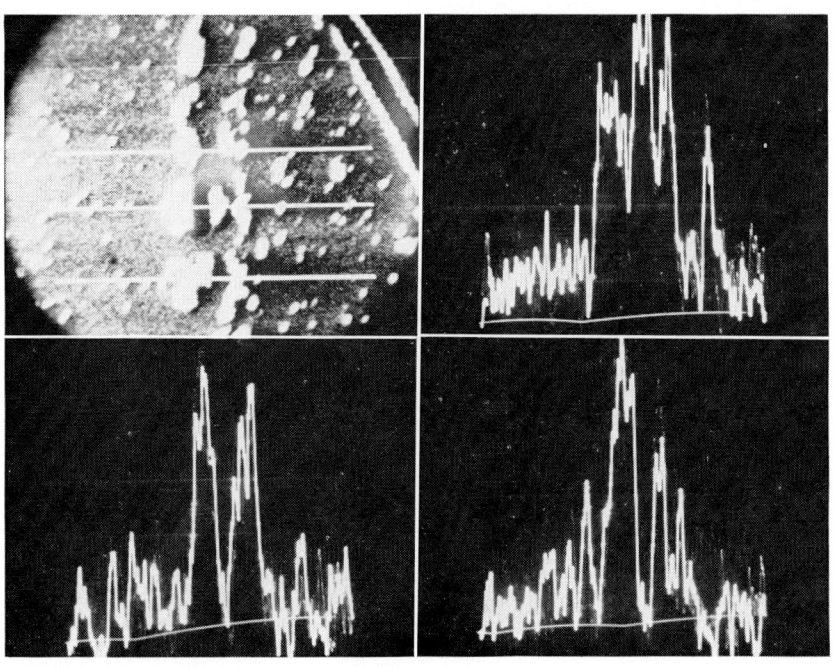

(bottom, left) Scan 1   (top, right) Scan 2   (bottom, right) Scan 3

Oscilloscope traces show W Lα x-ray intensity.

FIG. 4—Neighboring Carbides in the Electron Image (×500).

bides in a specimen of M-2 high-speed steel. In the electron image, white lines through the three groups of carbides indicate where linear scans were made with the microprobe. Oscilloscope traces represent the concentrations of iron (Fe), vanadium (V), chromium (Cr), molyb-

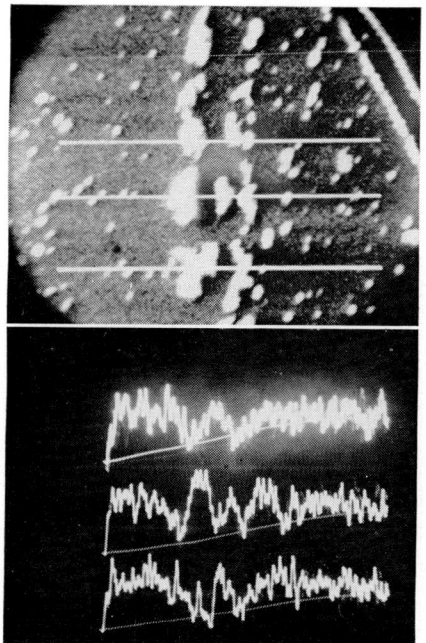

Oscilloscope traces show Cr Kα x-ray intensity.

Fig. 5—Neighboring Carbides in the Electron Image (×500).

denum (Mo), and tungsten (W). The Mo, W, and Cr concentrations vary only slightly from one carbide to another, but the Fe and V concentrations vary considerably.

The electron image of a large carbide is shown successively with Fe, V, W, and Cr x-ray images in Figs. 6 through 9. The oscilloscope traces show the element concentrations along the line through the center of the carbide. The

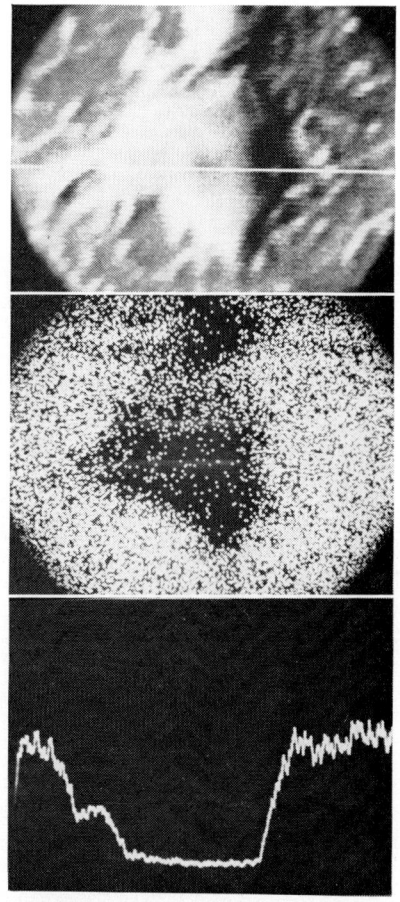

(*top*) Electron image showing a large MC carbide. The white line indicates where a line scan of the microprobe was made.

(*center*) Fe Kα x-ray image of the same area showing the Fe distribution.

(*bottom*) Oscilloscope trace showing Fe Kα x-ray intensity along the line of scan.

Fig. 6—Electron Image of a Large Carbide (×1500).

picture for Mo is not shown, but it was essentially the same as that for W.

The V concentration of the carbide was quite high, so it was evidently VC or MC. The illustrations indicate that the VC or MC carbide contains considerable amounts of the other alloying elements in solid solution.

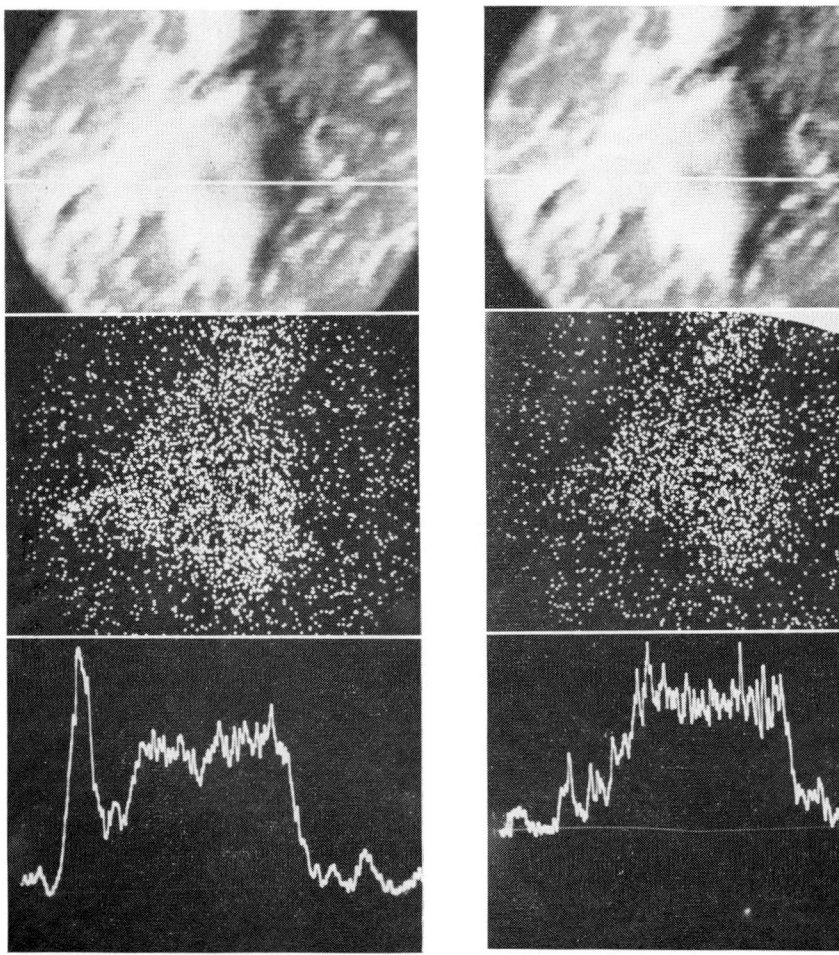

(top) Electron image showing a large MC carbide.
(center) V Kα x-ray image showing the V distribution.
(bottom) Oscilloscope trace showing V Kα x-ray intensity along the line scan.

Fig. 7—Electron Image of a Large Carbide (×1500).

(top) Electron image showing a large MC carbide.
(center) W Lα x-ray image showing the W distribution.
(bottom) Oscilloscope trace showing W Lα x-ray intensity along the line scan.

Fig. 8—Electron Image of a Large Carbide (×1500).

Figure 10 shows manganese sulfide (MnS) inclusions in a specimen of sulfur-bearing M-2 steel. The MnS inclusions appear as dark areas in the electron image. The line scan crossed three inclusions, and the oscilloscope traces show Mn and S x-ray intensities along the line. Variations in peak intensity are due to variations in inclusion size. The MnS enhances the machinability of this tool steel just as it does in regular free machining steel.

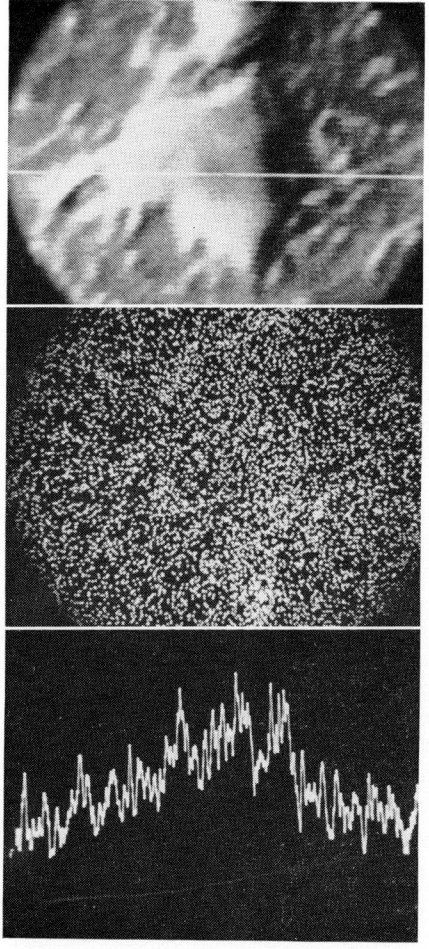

(*top*) Electron image showing a large MC carbide.
(*center*) Cr Kα x-ray image showing the Cr distribution.
(*bottom*) Oscilloscope trace showing Cr Kα x-ray intensity along the line scan.

FIG. 9—Electron Image of a Large Carbide (×1500).

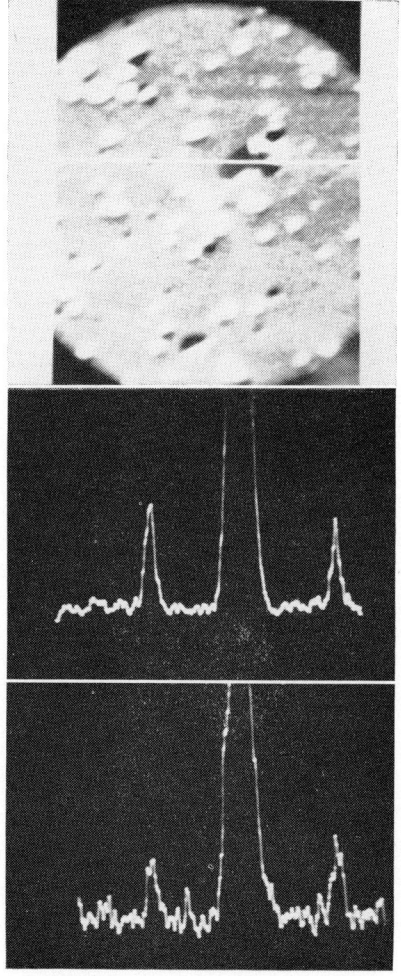

(*top*) Electron image showing MnS inclusions (dark areas).
(*center*) Oscilloscope trace showing Mn Kα x-ray intensity along the line scan.
(*bottom*) Oscilloscope trace showing S Kα x-ray intensity along the line scan.

FIG. 10—MnS Inclusions in a Specimen of Sulfur-Bearing M-2 Steel.

A carbide, identified as VC, was successively analyzed in the hardened, tempered, and annealed conditions to see if it changed composition during heat treatment. The tempering and annealing operations were conducted in vacuum using a sealed Vycor tube to make it possible to repolish lightly and probe the same carbide again. No significant changes in composition were observed in the center of the carbide.

Apparently it was massive enough so that the main body did not enter into the phase changes occurring during tempering and annealing.

Discussion

Some of the limitations of microprobe examination encountered in this investigation are related to (1) the size and distribution of carbides, and (2) the effective volume from which x-rays are emitted. These factors can make it difficult to obtain accurate, quantitative analyses of the matrix and of individual carbides.

It is difficult to analyze the matrix without interference from carbides even with an electron beam less than 1μ in diameter because the carbides are so close to each other. Furthermore, x-rays can be emitted by atoms 1 or 2μ below the surface. Thus the beam may appear to be in the matrix between carbides and still be exciting x-rays from a carbide just under the surface.

A carbide may be so thin that, when we attempt to analyze it, the matrix or another kind of carbide just below it can emit x-rays causing errors in the apparent analysis. A further limitation is that it is impossible to determine accurately the x-ray intensity from the numerous, tiny, carbide particles smaller in cross-section than the electron beam spot. In these cases only qualitative analyses can be made by employing linear scanning and x-ray imaging.

These problems, together with the inaccuracies introduced in making corrections for x-ray absorption and enhancement, have led to some uncertainty in identifying individual carbide particles. Selective electron-diffraction capability in the microprobe analyzer would help substantially to reduce this uncertainty by providing data on crystal structure to supplement chemical analyses.

Despite these difficulties the microprobe investigation of tool steel thus far has been enlightening, and we hope to learn more about the types of carbides and other phases present in tool steels.

VARIABLE BIAS ILLUMINATION CONTROL FOR THE ELECTRON MICROSCOPE

By J. O. McPartland[1] and J. L. Daniel[1]

Synopsis

Continuously variable control of illumination intensity in the electron microscope can be achieved by use of photoconductive cells for the grid bias resistance in a self-biased electron gun. The simple, highly stable device consists of several photo cells connected in series facing a control lamp, all positioned in a light-tight compartment. Changing the lamp brightness varies the grid resistance from 500,000 ohms to more than 30 megohms, controlling illuminating beam current over a wide range. Microscope stability is unaffected.

Continuously variable illumination intensity is of considerable value in electron microscopy, particularly in the laboratory where a wide variety of specimen types are examined. Requirements may range from minimum light levels to avoid specimen damage, minimize contamination, or aid in alignment, to maximum brightness for thick foils or replicas, or for reflection techniques. The electron beam intensity can be varied conveniently by control of the grid bias resistance in the self-biased electron gun. If photoconductive cells are used in place of the standard fixed or step resistance, the resistance can be continuously varied over a wide range by controlling light illumination falling on the photocells, as shown in Fig. 1.

A simple control device has been designed, in which the light-sensitive element consists of a group of five to seven photoconductive cells[2] connected in series and mounted in a clear oil-insoluble casting resin.[3] Suitable electrical plug connections are attached to the photoelement before mounting, to permit it to be substituted for the fixed bias resistor in the high voltage gun circuit shown in Fig. 2.

This photo-resistor is positioned in a light-tight compartment (for example, the high voltage oil tank) facing a six-volt control lamp mounted through the compartment cover. If the photo-resistor is submerged in oil, a lucite light pipe extends from the lamp to below the oil surface, to avoid surface reflection losses. The lamp is connected to the constant d-c supply shown in Fig. 3, with the ten-turn variable resistor mounted on the microscope control panel.

In operation, the grid bias resistance can be varied from 500,000 ohms to more than 30 megohms by varying the brightness of the control lamp through

---

[1] Engineer and senior scientist, respectively, Ceramics Research Operation, Hanford Laboratories, Richland, Wash.

[2] An example is the Clairex photoconductive cells, CL Type 602, peak = 5150 A, maximum voltage = 300, light resistance = 1 megohm; available from Allied Industrial Electronics Co., Chicago. Ill.

[3] A Titan boat resin with Catalyst P is available from Titan Chemicals, Inc., Seattle, Wash.

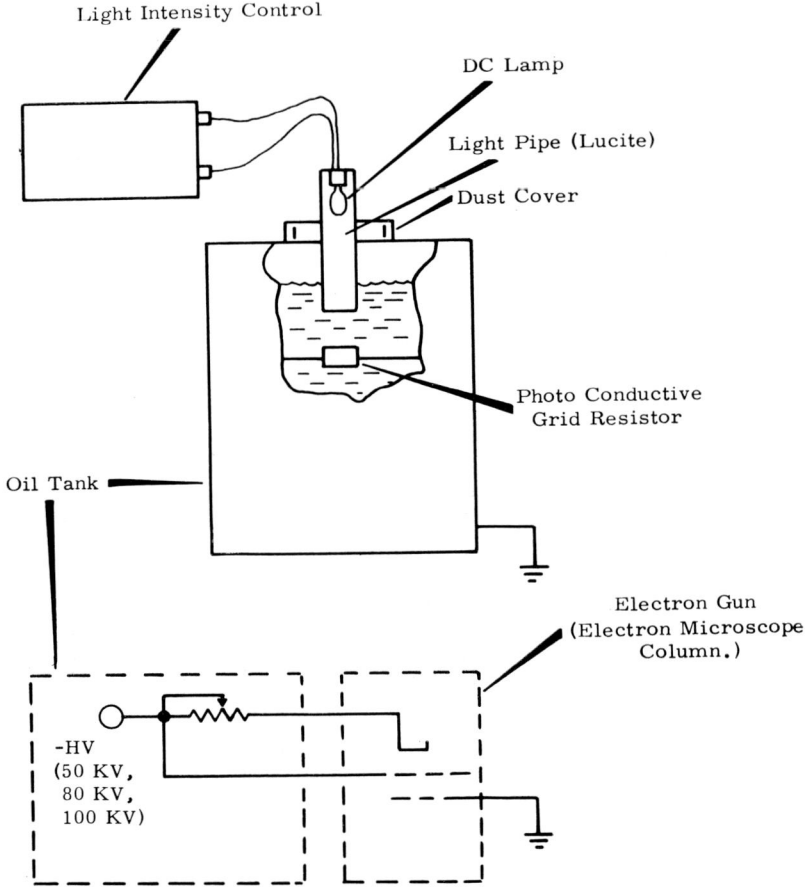

Fig. 1—Schematic Diagram of the Photoconductor Control of Electron Gun Grid Bias Resistance.

adjustment of the variable resistor in the lamp power supply. The lamp is operated at well below its rated value, resulting in a working lifetime of years of normal microscope use. The photocells also are long-lived, and have been found to give at least a year of use without maintenance. Since the photo-element is inexpensive and easy to make and install, occasional replacement is a simple procedure during routine microscope maintenance.

No difference in electron gun stability has been detected with the photo-bias in use. Instead, the added versatility of complete beam intensity control has greatly increased the operating convenience and over-all quality of performance of the microscope.

Acknowledgment:

The authors are indebted to Mr. R. W. Stephens for design of the constant d-c power supply.

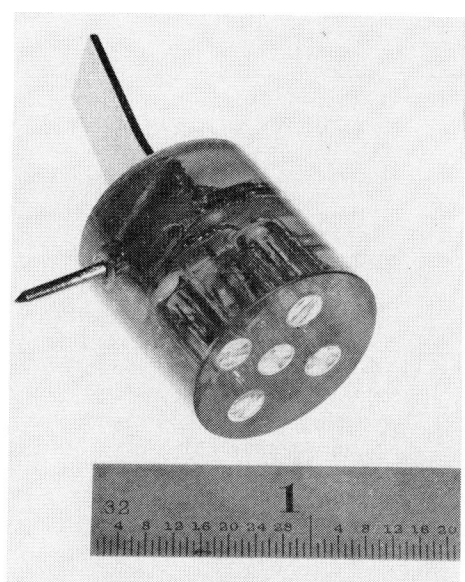

(*left*) a seven-cell element connected in series before mounting. (*right*) a five-cell element mounted in casting resin, ready for installation.

FIG. 2—Photoconductive Element.

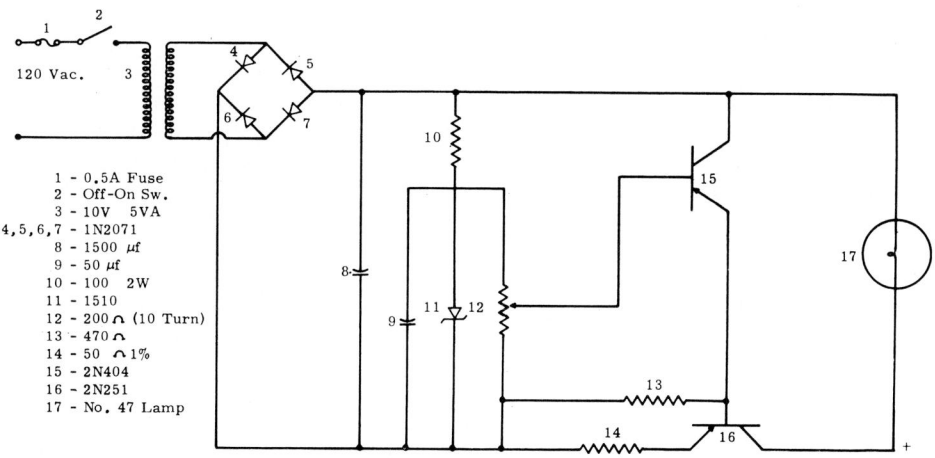

```
 1 - 0.5A Fuse
 2 - Off-On Sw.
 3 - 10V  5VA
4,5,6,7 - 1N2071
 8 - 1500 μf
 9 - 50 μf
10 - 100  2W
11 - 1510
12 - 200 Ω (10 Turn)
13 - 470 Ω
14 - 50 Ω 1%
15 - 2N404
16 - 2N251
17 - No. 47 Lamp
```

FIG. 3—Circuit Diagram of Constant Current Power Supply for Bias Control Lamp, 0 to 130 ma.

TWO NEW INDEXES TO THE POWDER DIFFRACTION FILE

By W. C. Bigelow[1] and J. V. Smith[2]

Synopsis

Two new indexes for use with the Powder Diffraction File have recently been published by ASTM. The Fink Index is a numerical index in book form that is based on the eight most intense lines of the powder patterns. It is designed to facilitate the identification of patterns that do not yield reliable relative intensity values, such as those generally obtained by electron diffraction and those sometimes obtained by X ray diffraction from specimens showing a high degree of preferred orientation. The Matthews Coordinate Index is an optical coincidence coordinate index published on Termatrex index cards. It has been developed to take advantage of modern data retrieval methods and to allow chemical information to be incorporated directly into the search procedure. The development and construction of both indexes are discussed in detail and several examples are given illustrating the use of both indexes in identifying typical unknown patterns.

In 1941 ASTM and the Joint Committee for Chemical Analysis by Powder Diffraction Methods[3] undertook, for the first time, the publication of the X-Ray Diffraction Data File,[4] a card file of powder diffraction data specifically designed for use in the identification of unknown crystalline materials by powder diffraction methods. Initially, this file was based on a compilation of data for about 1000 different compounds, and was organized according to an indexing system, which had been developed by Hanawalt, Rinn, and Frevel at the Dow Chemical Co. of Midland, Mich., and which had been described in the first three of a classic series of articles by these workers that laid the foundations for the effective use of powder diffraction techniques (1–7).[5] Over subsequent years, the data file has grown in size and improved in quality until it is now recognized throughout the world as *the* primary reference source for powder diffraction data. At present it contains data for approximately 13,000 compounds (about 6300 inorganic compounds and about an equal number of organic compounds), and it is currently planned to publish data for about 1500 new compounds each year.

The growth of the File to the point where it contained data for a large and

---

[1] Professor, Department of Chemical and Metallurgical Engineering, University of Michigan, Ann Arbor, Mich.

[2] Professor, Department of Geophysical Sciences, University of Chicago, Chicago, Ill.

[3] A cooperative activity of ASTM, the American Crystallographic Association, the British Institute of Physics, and the National Association of Corrosion Engineers.

[4] ASTM STP 48, available from the X-Ray Department of ASTM, Phila., Pa.

[5] The boldface numbers in parentheses refer to the list of references appended to this paper.

II - 1670

d	2.83	2.57	1.72	d λ=.708	$\frac{I}{I_1}$	d λ=.708	$\frac{I}{I_1}$
$\frac{I}{I_1}$	1.0	.8	.8	2.83 2.57 1.84 1.72 1.70 1.422 1.387 1.145 1.077 1.040	1.0 .8 .7 .8 .8 .8 .8 .7 .7 .7		
I	10	8	8				

PdS

Palladium Sulphide

(artificial Braggite)

Z =
$a_o =$ $b_o =$ $c_o =$
A =   C =
D =
n =   ω =   ε =

BM

II - 2171

d	2.57	1.72	2.83
$\frac{I}{I_1}$	.8	.8	1.0
I	8	8	10

PdS

Palladium Sulphide

II - 3404

d	1.72	2.83	2.57
$\frac{I}{I_1}$	.8	1.0	.8
I	8	10	8

PdS

Palladium Sulphide

FIG. 1—Three Data Cards for the Compound PdS from an Early Edition of the Powder Diffraction File.

continuously increasing number of compounds gave rise to problems involving effective means for using these data which have, in turn, prompted the Joint Committee to adopt several different indexing and data retrieval systems for use with the File. Earlier activities of the Joint Committee in this connection have been reviewed by W. L. Fink (8). The present paper describes the development, organization, and use of the Fink Index and the Matthews Coordinate Index,[6] which are the two indexes most recently developed by the editorial board. These indexes are particularly designed to provide, at a moderate cost, modernized indexing systems which permit greater speed, accuracy, and flexibility in the general use of the Powder Diffraction File, and which improve the effectiveness with which the File can be used in the identification of minor constituents of mixtures and in the identification of unknowns from their electron diffraction patterns. Since the principles upon which these new indexes are based differ considerably from those which formed the basis for previous indexes to the File, a brief description of the characteristics of the previous indexing system, and of the factors leading to the development of the new indexes, is given prior to describing the organization and use of the new indexes.

Development of the Indexes

The Hanawalt Index System:

From the time it was first published until 1963, all indexes to the Powder Diffraction File have been based primarily on the system developed by Hanawalt, Rinn, and Frevel at the Dow Chemical Co. (1–3). Under this system, each powder pattern was characterized by the interplanar spacings, d_1, d_2, and

---

[6] Both indexes are available from the X-Ray Department of ASTM, Phila., Pa. The Fink Index is designated as ASTM STP 48 M 3.

d_3, and the relative intensities, I_1, I_2, and I_3, of its three strongest lines. Initially the File contained three cards for each pattern, one for each of the following three arrangements of its three strongest lines: d_1, d_2, d_3; d_2, d_1, d_3; and d_3, d_1, d_2. The spacing range from above 10 A to below 1 A was divided into 87 intervals, chosen so that the number of the three strongest lines falling in each interval was approximately the same, and the cards were separated into 87 Hanawalt Groups corresponding to these intervals. Each card was assigned to the Hanawalt Group whose spacing interval included the first of these three d values. The cards were arranged in order within the groups according to the d values of their second lines. To identify an unknown, arrangements of three or more of the strongest lines of its diffraction pattern were formed, and the cards in the appropriate range (as determined by the second d value in each arrangement) of the corresponding Hanawalt Groups (determined by the first d value) were examined until a card was found whose three d values and relative intensities agreed satisfactorily with those of the unknown. All information listed on the card was then compared with the corresponding data for the unknown as the final test of the identification. Thus the data File initially served as its own index with respect to d values, although a separate index book, currently known as the Davey Index, listing the names and chemical formulas of the compounds in the file, was also provided. Figure 1 shows the set of three cards for the compound palladium sulfide (PdS) from an early edition of the File.

By the time the third set of data cards was published in 1950, the File contained more than 3000 different patterns and a total of nearly 10,000 cards, and it became time-consuming and inconvenient to search through the cards directly. The index book was therefore expanded to contain a numerical index in which the d values and relative intensities of the three arrangements of the three strongest lines were listed in Hanawalt Groups in the same order as they appeared on the cards into Hanawalt Groups) was subsequently abandoned, and the Hanawalt book index served as the principal approach to the data File for more than a decade. During this time, however, the number of patterns in the File has nearly quadrupled and the use of the File

2.94 - 2.90

	2.94	1.80	1.53	100	80	80	Be_3P_2	BERYLLIUM PHOSPHIDE	4-0669
	2.93	1.80	1.53	90	100	100	$(Zr,Ce)O_2$ (95% ZrO_2)	ZIRCONIUM, CERIUM OXIDE	2-1334
*	2.92	1.79	2.53	100	36	33	In_2O_3	INDIUM OXIDE	6-0416
	2.92	1.79	1.52	100	80	60	Li_2NH	LITHIUM AMIDE	6-0417
	2.90	1.78	2.96	90	100	90	Na_2MgAlF_7	SODIUM MAGNESIUM ALUMINUM FLUORIDE	5-0733
	2.91	1.78	2.52	100	80	70	$LiNH_2$	LITHIUM AMIDE	6-0418
	2.92	1.74	6.8	100	90	70	$Ca_4Si_3O_9(OH)_2$	FOSHAGITE	11-94
	2.92	1.74	3.46	100	60	50		VISEITE	5-0616
	2.92	1.72	1.49	100	100	100	CsCN	CESIUM CYANIDE	2-1442
	2.94	1.71	2.47	100	80	70	$FeO.WO_3$	FERBERITE	10-449
	2.94	1.71	2.24	100	90	80	$U_2V_6O_{21} \cdot 15H_2O$?	UVANITE	8-322
	2.93	1.71	1.45	100	80	80	$ZnNb_2O_6$	ZINC NIOBATE(V)	13-473
	2.93	1.70	2.47	100	60	50	$(Zn,Fe,Ca,Mn)WO_4$	SANMARTINITE	11-128
	2.92	1.70	2.45	100	70	50	Sr_2GeO_4	STRONTIUM GERMANATE	11-268
	2.94	1.70	1.46	100	100	85	$PbZrO_3$	LEAD ZIRCONATE	3-0655
	2.92	1.69	2.17	100	90	80	$FeNb_2O_6$	IRON NIOBATE	4-0672
	2.91	1.69	2.07	100	75	35	$\alpha-UH_3$	ALPHA URANIUM HYDRIDE	7-44
	2.92	1.69	2.06	100	90	80	Sr_2LaTaO_6	STRONTIUM LANTHANUM TANTALATE	11-574
	2.91	1.69	2.04	100	90	70	$Ca(ClO_2)_2$	CALCIUM CHLORITE	2-0758
	2.91	1.68	2.06	100	80	70	$BaSnO_3$	BARIUM STANNATE	3-0675
	2.91	1.68	2.05	100	40	20	$\beta-Th$ (AT. 1450°C)	BETA THORIUM	10-300
	2.90	1.54	1.13	100	70	70	Sm	SAMARIUM	6-0419
	2.92	1.49	8.3	80	100	60	$CuSiO_3 \cdot 2H_2O$	CHRYSOCOLLA	11-322
	2.92	1.27	1.24	60	100	100	Ni-U (42.5 WT. % Ni)	NICKEL URANIUM	9-282
	2.93	1.23	0.91	100	100	100	TiTe	TITANIUM TELLURIDE	6-0420
	2.91	1.21	1.81	100	100	80	$TiTe_2$	TITANIUM DI TELLURIDE	6-0421

FIG. 2—Portion of a Typical Page From the Hanawalt Index Book.

cards in the File, together with the chemical formula of the compound and the number of the card in the File as shown in Fig. 2. Since a preliminary search based on the three strongest lines could be made so much more rapidly and easily in this index book than in the data File itself, the index feature of the File (which included the publication of three cards per pattern and the arrangement of has expanded into many new areas, so that this index no longer fulfills the requirements of many users as well as might be desired.

Development of the Fink Index:

Particular difficulty has been encountered in attempting to use the Hanawalt index for the identification of unknowns by electron diffraction methods, since the interplanar spacings and relative intensities of the diffraction lines cannot generally be determined as accurately from the electron diffraction patterns as from X ray diffraction patterns. Interest in the practical applications of electron diffraction techniques increased steadily during the 1950's as the number of modern electron microscopes equipped for electron diffraction work increased in laboratories engaged in nonbiological research. In 1960, following a national survey to determine the level of interest in electron diffraction techniques, and a series of discussions by several scientific organizations to consider the problems involved in identifying unknown materials from electron diffraction patterns, the Joint Committee agreed to undertake a study of the problems involved in electron diffraction work. Its ultimate objective was to devise a method of facilitating the identification of unknown materials from powder electron diffraction patterns (9).

At the outset of this study it was the opinion of most persons directly concerned with the matter that it would be desirable to establish a new compilation of electron diffraction powder data independent of, but similar to, the existing file of X ray data. The study immediately showed that this would be a long and difficult task that could not provide a practical solution to the problem at hand in the foreseeable future. In seeking a solution that could be more readily accomplished, electron diffraction data contributed by a large number of cooperating individuals and organizations were carefully compared with X ray data for corresponding compounds from the File. This comparison showed that, in a majority of the cases, the d values reported for the electron diffraction patterns agreed with those reported for the X ray patterns within about one per cent, and that the magnitude of the difference depended largely on the care used in calibrating and measuring the electron diffraction patterns. It was also observed that the same group of lines had high intensities (that is, would be assigned a value of I/I_1 greater than about 30, or would be qualitatively rated as being of medium intensity or stronger) in both the X ray and electron diffraction patterns of a given material, although the order of their ranking with respect to intensity might differ considerably for the two types of patterns, depending on the nature of the specimens and the experimental conditions used in taking the patterns in each case. Thus it appeared that the difficulties encountered in using the Hanawalt index for identifying electron diffraction patterns resulted principally from the differences which frequently occurred in the relative intensities of corresponding lines in X ray and electron diffraction patterns of the same material. Under the Hanawalt system, the interplanar spacing, d_1, of the single strongest line of the reference pattern determines either the Hanawalt group in which the pattern is listed, or the location of the pattern within a group, for all three listings of the pattern in the index. Failure to select the correct line of the unknown pattern as the strongest line for use in searching the index, because of such intensity differences, often makes it difficult to effect an identification using the Hanawalt system. Similar difficulties are also encountered in dealing with X ray patterns from mixtures (where coincidence of lines from different components produces anomalous intensities) and from specimens having preferred orientations of the crystallites.

These results clearly indicated that it would be more effective to continue using the existing file of X ray powder diffraction data for the identification of electron diffraction patterns than to attempt to compile a new file of electron diffraction data for this purpose. However, it appeared desirable to attempt to develop a new index to the file which would accommodate uncertainties in d values and relative intensities more readily than the existing index. It was also concluded that the new index should (1) be based on more than three d values per reference pattern, (2) contain more than three entries for each reference pattern, and (3) list the entries in an order not directly dependent on relative intensities.

In 1960 and 1961 a preliminary index of this type was prepared for the inorganic compounds in the File using the d values of the eight strongest lines of each pattern but no relative intensity data. This preliminary index was evaluated by a number of persons and found to be more satisfactory in dealing with electron and X ray diffraction patterns which did not yield reliable intensity data than the Hanawalt index. It also appeared that the new index could be used effectively in identifying with a variety of other types of patterns, provided appropriate search procedures were used. In addition, the listing of eight d values for each pattern in the new index was found to permit a fairly critical comparison of the unknown and reference patterns in most cases, thereby reducing considerably the number of patterns that needed to be looked up in the File, and markedly speeding up the initial stages of the search.

In view of its success in dealing with electron diffraction patterns and the strong indications that the new index was suited for general use in the identification of all types of powder patterns, the Joint Committee decided to publish an index book based on this system, and to name it in honor of Dr. W. L. Fink who, as Chairman of the Joint Committee, had fostered its development. Minor modifications were made in the format and organization of the preliminary index to improve its flexibility and versatility, and to facilitate the use of computer techniques in compiling and processing the data, and the first edition of the Fink index was published during the summer of 1963.

In addition to members of the Joint Committee and its editorial board, there were many others who contributed to the development of this index. Of these, the following deserve particular mention: Karl E. Beu, of the Goodyear Atomic Corp., Piketown, Ohio; John R. Dorsey of the National Security Agency, Fort George G. Meade, Md.; and Dr. Victor Hicks of Remington Rand Univac, St. Paul, Minn., who were particularly active in organizing the initial phases of the studies of the electron diffraction problem; Professor W. C. Bigelow, consultant to the Joint Committee, who carried out the studies of the problems involved in the identification of electron diffraction patterns and evaluated the performance of the preliminary version of the index in dealing with electron diffraction data; Professor J. V. Smith, editor of the powder diffraction data file, who developed the details of the format of the index and worked out procedures for using punched card data processing techniques in the preparation of the index; H. McMurdie and H. Swanson of the National Bureau of Standards, Washington, D. C., who were in charge of preparing the master set of punched cards; Dr. L. E. Kuentzel of the Wyandotte Chemical Corp., Wyandotte, Mich., and his daughter, who carried out the processing of the data cards, and the Wyandotte Chemical Corp. which provided the necessary facilities for this operation; and Mr. A. S. Beward, assistant editor of the Diffraction Data File, who was in charge of the technical processes associated with preparation and publication of the index.

Development of the Matthews Index:

Whereas the Fink index was developed primarily to meet a specific need relating to the use of the File, the Matthews index was largely the result of a long-standing desire on the part of the Joint Committee and others interested in the File to provide a mechanical or semi-mechanical search method based on modern techniques of data processing and information retrieval, which would be more rapid, accurate, convenient, and interesting to use than the book-type indexes, and which could be expanded to permit information on the physical properties and chemical composition of the unknown to be incorporated directly and effectively into the search procedure. Two previous attempts had been made to achieve these goals. Beginning in 1950, special editions of the File were offered on keysort cards which could be edge-punched for chemical and physical properties in addition to the usual d value and relative intensities, and which made it possible to take advantage of the keysort data retrieval system in searching the File. In 1952 the File was published on IBM cards to allow IBM machine-sorting and data-processing techniques to be applied in using the File. The details of these undertakings, which were only moderately successful, have been reviewed elsewhere (8).

In 1957 Dr. F. W. Matthews of the Central Research Laboratory of Canadian Industries, Ltd., McMasterville, Quebec, Canada, presented a talk at the meeting of the International Union of Crystallography in Cambridge, England, in which he described the adaptation of the optical coincidence-coordinate index system for use with the File. This index system fulfilled all of the objectives cited above, and in addition provided for the first time, a convenient method for selecting from the file all entries having a specified set of characteristics in common. In subsequent years, Dr. Matthews made an extensive evaluation of the system, using an experimental index on punched IBM cards covering the minerals contained in the first nine sets of data cards (10). This work clearly demonstrated the merits of the coordinate indexing system in working with the File; however, the publication for sale of such an index covering any substantial portion of the File was not feasible until 1961, when Jonkers Business Machines, Inc., of Gaithersburg, Md., developed their Termatrex coordinate index cards having the capacity to accommodate data for as many as 10,000 patterns. With the development of these cards, the editorial board published in 1962 a preliminary edition of a coordinate index covering the patterns of the inorganic compounds in the first twelve sets of data cards. Modifications were made in this index on the basis of evaluations which were made in a number of laboratories, and the first edition of it was published in the spring of 1963 and named in honor of Dr. Matthews in recognition of his work leading to its development. Several other persons have also contributed to this development, including particularly: Dr. Smith, who assisted Dr. Matthews in developing the system and procedures used in organizing and coding the data for listing on the coordinate cards, and Mr. Beward, who carried out the coding procedures; Mr. Freeman H. Dyke of Jonkers Business Machines, Inc. designed the coordinate cards for the index and took charge of listing the data on the cards. The index is published for the Joint Committee by Jonkers Business Machines, Inc., using their Termatrex data processing system and equipment.

Probable Future Developments:

At present both the Matthews and Fink indexes cover only inorganic compounds in the File, although consideration is being given to extending them to include organic compounds in the near future. With the Matthews index it is possible to prepare special cards for any type of information. At present this index includes special cards for the presence of alloys, minerals, and hydrates. In the future, cards may be issued covering other physical properties of general interest, and facilities may become available which will make it possible to prepare, as a special extra-cost service, cards covering properties or characteristics of a less general nature that are of particular interest to a limited number of users. The Joint Committee has recently engaged Dr. Vladimir Vand of the Pennsylvania State University as a consultant, responsible for improving the present indexes and developing new ones. One such development that is being given serious consideration is the establishment of a comprehensive index to the File in a computing facility, so that extensive search procedures or data compilations can be undertaken as special services for users requiring them. In the meantime, it is the intent of the editorial board to continue its present and past policies of attempting to insure that the File and its existing indexes are as accurate and convenient to use as possible by updating and revising them, as appears necessary, at approximately yearly intervals when new sets of cards are published for addition to the File.

CHARACTERISTICS AND USE OF THE FINK INDEX

As indicated above, the Fink index to the Powder Diffraction File has been developed primarily for the purpose of facilitating the identification of powder diffraction patterns that do not provide reliable values for the relative intensities of the diffraction lines, such as those frequently obtained in electron diffraction work and in X ray diffraction studies of mixtures and of specimens having preferred orientations of the crystallites. It is fully expected, however, that this index will ultimately find wide use as a general-purpose index for identification of all types of patterns. The Fink index is published in book form and thus affords the advantages of low cost, portability, and ease and convenience of use, without requiring expensive auxiliary equipment.

Organization of the Fink Index:

The Fink index uses the d values of the eight strongest lines less than 9.99 A in spacing to characterize each pattern

TABLE 1—ORDER OF LISTING OF d VALUES IN THE EIGHT ENTRIES FOR EACH COMPOUND IN THE FINK INDEX.

Entry	Order of listing of d Values
1	$d_1, d_2, d_3, d_4, d_5, d_6, d_7, d_8$
2	$d_2, d_3, d_4, d_5, d_6, d_7, d_8, d_1$
3	$d_3, d_4, d_5, d_6, d_7, d_8, d_1, d_2$
4	$d_4, d_5, d_6, d_7, d_8, d_1, d_2, d_3$
5	$d_5, d_6, d_7, d_8, d_1, d_2, d_3, d_4$
6	$d_6, d_7, d_8, d_1, d_2, d_3, d_4, d_5$
7	$d_7, d_8, d_1, d_2, d_3, d_4, d_5, d_6$
8	$d_8, d_1, d_2, d_3, d_4, d_5, d_6, d_7$

contained in the File. No relative intensity values are listed in the index, nor are they used in determining the order of listing of the data or the patterns in the index. Values of d greater than 10 A are used only if their $I/I_1 = 100$. When it is necessary to choose among several d values having the same relative intensities, selection is made in decreasing numerical order. In cases where fewer than eight d values are given on the data cards, all the d values given are used, and the remainder of the eight are listed as 0.00 in the index.

All eight of the d values used to characterize each pattern are entered eight different places in the index. For the first entry of each pattern, the eight d values are arranged in decreasing numeri-

TABLE 2—d VALUE INTERVALS FOR GROUPING OF ENTRIES IN THE FINK INDEX.

Group	Interval	Range	Group	Interval	Range
1	10.00 and greater		52	2.67–2.66	0.01
2	9.99–8.00	1.99	53	2.65–2.64	0.01
3	7.99–7.50	0.49	54	2.63–2.62	0.01
4	7.49–7.00	0.49	55	2.61–2.60	0.01
5	6.99–6.50	0.49	56	2.59–2.58	0.01
6	6.49–6.00	0.49	57	2.57–2.56	0.01
7	5.99–5.80	0.19	58	2.55–2.54	0.01
8	5.79–5.60	0.19	59	2.53–2.52	0.01
9	5.59–5.40	0.19	60	2.51–2.50	0.01
10	5.39–5.20	0.19	61	2.49–2.48	0.01
11	5.19–5.00	0.19	62	2.47–2.46	0.01
12	4.99–4.80	0.19	63	2.45–2.44	0.01
13	4.79–4.60	0.19	64	2.43–2.42	0.01
14	4.59–4.40	0.19	65	2.41–2.40	0.01
15	4.39–4.30	0.09	66	2.39–2.38	0.01
16	4.29–4.20	0.09	67	2.37–2.36	0.01
17	4.19–4.10	0.09	68	2.35–2.34	0.01
18	4.09–4.00	0.09	69	2.33–2.32	0.01
19	3.99–3.90	0.09	70	2.31–2.30	0.01
20	3.89–3.80	0.09	71	2.29–2.28	0.01
21	3.79–3.70	0.09	72	2.27–2.26	0.01
22	3.69–3.65	0.04	73	2.25–2.24	0.01
23	3.64–3.60	0.04	74	2.23–2.22	0.01
24	3.59–3.55	0.04	75	2.21–2.20	0.01
25	3.54–3.50	0.04	76	2.19–2.18	0.01
26	3.49–3.45	0.04	77	2.17–2.16	0.01
27	3.44–3.40	0.04	78	2.15–2.14	0.01
28	3.39–3.35	0.04	79	2.13–2.12	0.01
29	3.34–3.30	0.04	80	2.11–2.10	0.01
30	3.29–3.25	0.04	81	2.09–2.08	0.01
31	3.24–3.20	0.04	82	2.07–2.06	0.01
32	3.19–3.15	0.04	83	2.05–2.04	0.01
33	3.14–3.10	0.04	84	2.03–2.02	0.01
34	3.09–3.05	0.04	85	2.01–2.00	0.01
35	3.04–3.00	0.04	86	1.99–1.95	0.04
36	2.99–2.98	0.01	87	1.94–1.90	0.04
37	2.97–2.96	0.01	88	1.89–1.85	0.04
38	2.95–2.94	0.01	89	1.84–1.80	0.04
39	2.93–2.92	0.01	90	1.79–1.75	0.04
40	2.91–2.90	0.01	91	1.74–1.70	0.04
41	2.89–2.88	0.01	92	1.69–1.65	0.04
42	2.87–2.86	0.01	93	1.64–1.60	0.04
43	2.85–2.84	0.01	94	1.59–1.55	0.04
44	2.83–2.82	0.01	95	1.54–1.50	0.04
45	2.81–2.80	0.01	96	1.49–1.40	0.09
46	2.79–2.78	0.01	97	1.39–1.30	0.09
47	2.77–2.76	0.01	98	1.29–1.20	0.09
48	2.75–2.74	0.01	99	1.19–1.10	0.09
49	2.73–2.72	0.01	100	1.09–1.00	0.09
50	2.71–2.70	0.01	101	0.99 and less	
51	2.69–2.68	0.01			

cal order (*left* to *right*). Cyclic permutation is then used to establish the order in the remaining seven entries, as shown in Table 1 with $d_1 > d_2 > d_3 > \cdots > d_8$. To facilitate searching the index on the basis of two or more d values simultaneously, the d value range from above 10 A to below 1.0 A is divided into 101 intervals as shown in Table 2, and the entries are separated into 101 correspond-

3.04-3.00

3.00	2.48	2.25	2.04	3.84	3.63	3.41	3.11	PHOSPHOROUS ACID	3 297
3.04	2.47	1.66	4.99	4.68	3.95	3.32	3.28	TAVORITE BASIC LITHIUM IRON PHOSPHATE	10 424
3.02	2.47	2.13	1.91	1.74	1.42	1.35	4.28	BARIUM HEXA BORIDE	11 213
3.02	2.47	2.34	2.20	2.14	2.10	2.00	3.23	TUNGSTEN SILICON	8 350
3.03	2.46	1.92	1.76	1.74	1.24	1.03	0.83	COOPERITE PT PD NI SULFIDE	9 287
3.03	2.46	2.16	1.77	7.35	5.32	4.31	3.07	DELTA SODIUM FLUOTHORATE	12 552
3.02	2.46	1.91	1.75	1.73	1.50	1.24	1.16	COOPERITE PLATINUM SULFIDE	2 642
3.02	2.46	2.11	1.87	1.73	1.36	4.24	3.81	POTASSIUM ACETYLIDE	4 646
3.02	2.46	2.14	1.51	1.29	1.23	0.98	0.95	CUPRITE COPPER OXIDE	5 667
3.01	2.46	2.20	1.81	1.72	1.70	1.52	1.38	CHROMIUM URANIUM CARBIDE	12 464
3.01	2.46	2.13	1.90	1.74	1.29	0.84	4.25	NEPTUNIUM ALUMINUM 1 TO 3	6 542
3.01	2.45	2.40	2.09	1.48	8.41	3.45	3.32	ACID AMMONIUM COBALTIMOLYBDATE 3 HYDRATE	4 647
3.00	2.45	2.11	1.72	1.63	1.49	1.34	4.95	POTASSIUM FLUOALUMINATE	3 615
3.02	2.44	2.41	2.21	2.13	1.52	1.41	3.21	TANTALUM SILICIDE 5 TO 3	6 594
3.02	2.44	1.96	1.82	1.49	4.80	4.05	3.51	HAGGITE VANADIUM HYDROGEN OXIDE	13 163
3.04	2.43	6.99	4.84	3.61	3.42	3.33	3.21	ZINCIAN ROCKBRIDGEITE	6 270
3.02	2.43	2.24	1.95	5.40	5.20	3.27	3.15	PHOSPHORUS PENTA OXIDE	1 213
3.01	2.41	1.97	1.66	1.52	10.6	9.40	4.80	BISMUTH OSMATE	1 275
3.01	2.41	1.97	1.66	1.52	9.40	6.20	4.80	BISMUTH OSMATE	1 275
3.00	2.41	2.12	1.88	1.46	1.28	0.95	4.04	MERCURY TITANIUM 1 TO 1	7 117
3.00	2.41	1.87	1.66	1.49	1.27	1.26	3.40	BASIC TELLURIUM NITRATE	1 862
3.04	2.40	1.83	1.38	1.36	9.14	4.57	3.34	PYROPHYLLITE ALUMINUM SILICATE HYDRATE	2 613
3.04	2.40	2.36	2.02	1.86	3.47	3.24	3.07	NIOBIUM IODIDE	12 93
3.00	2.40	2.14	1.88	1.87	1.78	3.18	3.04	MERCURY PEROXIDE	13 140
3.04	2.39	2.03	10.2	5.47	5.26	3.88	3.38	CYANOTRICHITE BASIC COPPER AL SULFATE HYD	11 131

3.04-3.00

2.47-2.46

2.47	2.14	1.52	1.29	0.98	0.96	0.88	0.83	VANADIUM NITRIDE	2 1064
2.47	2.14	2.10	1.46	1.34	1.24	1.17	2.91	NEODYMIUM NICKEL 1 TO 5	12 501
2.47	2.14	2.10	1.46	1.35	1.26	1.17	2.91	NICKEL PRASEODYMIUM 5 TO 1	12 502
2.47	2.14	2.05	1.83	1.76	3.48	3.01	2.70	HIGH BORACITE MAGNESIUM BORATE CHLORIDE	3 1034
2.46	2.14	1.51	1.29	1.23	0.98	0.95	3.02	CUPRITE COPPER OXIDE	5 667
2.46	2.14	2.16	1.44	3.06	2.98	2.66		SULFUR DI OXIDE	5 428
2.46	2.14	1.98	6.48	5.56	3.36	3.11	2.77	POTASSIUM IODORHENATE	9 49
2.47	2.13	1.91	1.74	1.42	1.35	4.28	3.02	BARIUM HEXA BORIDE	11 213
2.46	2.13	1.51	1.28	1.23	1.06	0.98	0.95	COBALT OXIDE	9 402
2.46	2.13	1.90	1.74	1.29	0.84	4.25	3.01	NEPTUNIUM ALUMINUM 1 TO 3	6 542
2.46	2.13	1.95	1.91	4.41	4.20	3.33	3.05	CALCIUM THIOCYANATE 3 HYDRATE	1 829
2.47	2.12	2.10	1.61	1.44	1.42	4.01	2.64	NEODYMIUM CHLORIDE	13 112
2.47	2.12	1.99	1.93	1.77	7.72	3.23	2.77	LEAD ALUMINUM SILICATE	3 489
2.46	2.12	1.83	1.56	1.06	3.69	3.00	2.61	TETRAHEDRITE CU.FE.ZN SB SULFIDE	11 107
2.46	2.12	1.93	1.87	4.05	3.14	2.84	2.48	LOW OR ALPHA CRISTOBALITE SI DI OXIDE	11 695
2.46	2.12	3.29	3.17	2.98	2.88	2.54	2.52	CLINOENSTATITE MAGNESIUM META SILICATE	13 415
2.46	2.11	2.01	1.61	1.57	1.44	3.97	3.52	RUBIDIUM BOROHYDRIDE	8 224
2.46	2.11	1.87	1.73	1.36	4.24	3.81	3.02	POTASSIUM ACETYLIDE	4 646
2.46	2.11	1.75	1.52	7.06	4.54	3.53	2.63	CHAMOSITE BASIC IRON IC MG AL SI OXIDE	7 329
2.47	2.10	1.74	1.60	1.42	1.34	0.93	4.03	STRONTIUM CHLORIDE	6 537
2.47	2.10	1.80	1.57	5.53	3.80	3.13	2.91	STRONTIUM ALUMINUM SILICATE	10 19
2.46	2.10	2.01	1.56	1.43	1.16	4.02	3.49	SILVER PERCHLORATE .AT 180 DEGREES C.	2 401
2.46	2.10	1.93	1.74	1.48	1.45	4.02	2.84	SCANDIUM OXIDE	5 629
2.46	2.10	2.01	1.43	1.34	1.24	0.91	0.81	CHROMIUM NIOBIUM	5 701
2.46	2.10	1.62	1.44	1.38	1.21	1.04	2.64	ANTIMONY TITANIUM 2 TO 1	6 519
2.46	2.10	2.00	1.57	1.53	1.38	1.30	2.62	CALCIUM FLUORIDE.ALUMINUM OXIDE	6 717

2.47-2.46

(*top*) First entry for cuprite in Group 35.
(*bottom*) Second entry for cuprite in Group 62.

FIG. 3—Portions of Pages From the Fink Index Showing the First Two Entries for Cuprite, Cu_2O.

ing groups by assigning each entry to the group whose interval includes the first d value (counting from the left) of the entry. Entries within each group are arranged in numerical order of their second d values. The d value interval for the appropriate group is printed on every page of the index book. In addition to the eight d values, each entry in the index includes the mineral name and chemical name of the substance (with the latter shortened if necessary to fit into the available space) and the number of the card in the File from which the data were taken. For example, the first entry for cuprite occurs in the 35th Fink group, whose d-value interval extends from 3.00 to 3.04 A, between the third and first entries, respectively, for potassium acetylide and chromium uranium carbide. The second entry for cuprite appears in Fink group No. 62, whose interval extends from 2.46 to 2.47 A, between the fourth and fifth entries, respectively, for magnesium borate fluoride and sulfur dioxide. Portions of the pages of the index on which these two entries occur are shown in Fig. 3.

Use of the Fink Index:

The use of the Fink index in identifying unknown materials from their powder diffraction patterns is relatively simple and straightforward; however, the following points regarding the procedures used in selecting and arranging the data in the index must be carefully noted (particularly by persons long accustomed to using the Hanawalt index) if an effective search procedure is to be achieved:

1. The d values listed include not only those for the three strongest lines of the pattern, but also those for the next five strongest lines.

2. No information is given concerning the relative intensities of these lines or their rank with regard to intensity, although in most cases at least half of them will have relative intensities greater than 30.

3. The arrangement of the data, including the grouping of the entries, the ordering of the entries within the groups, and the listing of the d values for each entry, is determined strictly by the numerical magnitudes of the d values, and is entirely independent of their relative intensities.

In general, the design and execution of an identification procedure will include the following steps:

1. Note all available information concerning the chemical composition of the unknown.

2. Note, in descending numerical order, the d values of the high intensity lines of the unknown diffraction pattern (that is, those lines with I/I_1 greater than 30, or those that would be qualitatively rated of medium intensity or stronger).

3. Chose an appropriate one of these, d_s, and the next smaller one, d_{s-1}, for use in starting the search. This choice critically influences the effectiveness of the

TABLE 3—X-RAY DIFFRACTION DATA FOR PRECIPITATES FROM CHROMIUM-NICKEL-STAINLESS STEEL.

d, A	I/I_1
2.30	m[a]
2.23	w
2.15	w
2.13	m[a]
2.03	s[a]
1.97	m[a]
1.84	w
1.80	wm[a]
1.76	w
1.70	w
1.62	w
1.45	m[a]
1.35	m[a]
1.30	w
1.27	w
1.22	wm[a]
1.19	s[a]
1.18	w
1.17	ms[a]

[a] High intensity lines chosen for use in identification procedure.

search procedure, and is discussed in greater detail below.

4. Enter the Fink group whose d value interval includes d_s, making suitable allowance for experimental uncertainties. Examine the second column of the index to find the section of this group which extends from 1 to 3 per cent above and below the value of d_{s-1}. Examine the data for the patterns listed in this section of the group for agreement with the unknown pattern to effect a preliminary identification.

5. If a preliminary identification is achieved, compare the data for the un-

known pattern with the data on the corresponding card in the File to achieve a final identification.

6. If a satisfactory identification is not achieved, compare the unknown pattern successively with data in sections of the same group, which include the d values d_{s-2}, d_{s-3}, etc., for the successively smaller d values selected in step two.

7. If a satisfactory identification is still not achieved, make a new choice of d_s and repeat the process in steps three, four, five, and six.

Four examples illustrating this procedure are given below.

Example 1—An X ray diffraction pattern obtained from precipitate particles separated from a chromium-nickel stainless steel gave the data listed in Table 3.

The d values of the high intensity lines which would be chosen according to step two for use in the search procedure are indicated in the Table. There are two logical choices for d_s. The simplest and most straightforward approach is to choose the largest of the high intensity lines as d_s, giving: $d_s = 2.30$ and $d_{s-1} = 2.13$, $d_{s-2} = 2.03$, $d_{s-3} = 1.97$, etc. In this case, enter groups 68, 69, 70, and 71 (allowing for about 2 per cent experimental uncertainty) searching those sections where the entries contain values extending from about 2.15 to 1.90 in the second column, keeping in mind that the unknown is most likely a carbide, oxide, or nitride of chromium, iron, or some other metal commonly found in stainless steels. Satisfactory preliminary agreement should be noted between the unknown pattern and the following entry in group 71: 2.28 2.12 2.04 1.96 1.81 1.35 1.19 1.17 CHROMIUM CARBIDE 11-550 and reference to data card 11-550 would confirm the unknown as chromium carbide (Cr_7C_3).

An alternate approach would be to choose $d_s = 2.03$, since this is one of the two strongest lines of the unknown pattern and is most likely to correspond to a high intensity line in the reference pattern. In this case, $d_{s-1} = 1.97$, $d_{s-2} = 1.80$, $d_{s-3} = 1.45$, etc. Searching would then be carried out in groups 82, 83, 84, 85, and 86 (again allowing for about 2 per cent experimental uncertainty) in the sections where d values ranging from about 2.00 to 1.95 A appear in the second column of the entries. With careful searching, the agreement between the unknown pattern and the entry: 2.04 1.96 1.81 1.35 1.19 1.17 2.28 2.12 CHROMIUM CARBIDE 11-550 should be recognized, again leading to the identification of the unknown as Cr_7C_3.

It is particularly important to note that, in this second case, the d values of the 1.19 and 1.17 lines are not used for d_{s-1} and d_{s-2} even though they have higher intensities than the 1.97 line, because the d values of the reference patterns are listed in numerical sequence in the Fink index, not in order of relative intensity. For the same reason, the search procedure was equally successful when employing a line of medium intensity for d_s as when one of the strongest lines was used. It would be particularly undesirable to choose $d_s = 1.19$. Since this has one of the smallest numerical values of any of the high intensity lines of the unknown pattern, any entry based on the corresponding line of the reference pattern would probably consist of an awkward sequence of d values that would be less likely to be anticipated and recognized in the search procedure than would entries based on the larger d values. The choice of one of the smallest d values as the basis for the search procedure will usually lead to particularly serious difficulties in dealing with unknown patterns from mixtures of phases. As a general rule, therefore, priority should be given to using one of the largest of the d values for the high intensity lines of the unknown pattern as the basis of the search procedure. There is no particular advantage in choosing the line of highest intensity if it does not fulfill this requirement; on the other hand, choice of the highest intensity line may involve serious inconveniences and disadvantages if it has one of the smallest d values.

Example 2—A good pattern of an unknown mineral, for which no chemical data were available, gave the X ray diffraction data listed in Table 4. The high intensity lines indicated were chosen for use in searching the Fink index and suggest starting with one of two alternate search procedures, based on the following sequences of these lines:

Procedure 1	Procedure 2
$d_s = 5.2$ ms	$d_s = 2.77$ s
$d_{s-1} = 3.90$ ms	$d_{s-1} = 2.51$ s
$d_{s-2} = 3.50$ m	$d_{s-2} = 2.46$ s
$d_{s-3} = 2.77$ s	$d_{s-3} = 2.35$ m
$d_{s-4} = 2.51$ s	$d_{s-4} = 2.27$ ms
$d_{s-5} = 2.46$ s	$d_{s-5} = 2.25$ ms

The relative merits of these two procedures cannot be predicted, except for the case where the unknown is a single phase, when it would probably be more efficient to use Procedure 2 based on the characteristic grouping of three adjacent lines of strong intensity. It is, of course, also possible to look for this characteristic grouping in Procedure 1, in which case the 2.77 line would be expected in either the third or fourth column of the index book.

For Procedure 1, searching should be carried out in those portions of groups 10 and 11 (corresponding to a range of from 5.00 to 5.39 A for d_s) where d values ranging from 3.8 to 4.0 A (corresponding to an experimental error of ± 3 per cent) appear, looking for d values near 3.50 or 2.77 in the third and fourth columns of the index book.

Because the specimen is known to be a mineral, only those entries with mineral names need be considered. By this approach the following entries showing satisfactory agreement with the data for the unknown pattern are found in group 11: 5.13 3.89 2.77 2.52 2.46 2.27 2.25 1.75 CHRYSOLITE MAGNESIUM ORTHO SILICATE 7-156; 5.13 3.89 2.77 2.52 2.46 2.27 2.25 1.75

TABLE 4—DATA FOR IDENTIFICATION OF UNKNOWN MINERAL OF EXAMPLE 1 ILLUSTRATING USE OF FINK INDEX.

Unknown Pattern		Forsterite, 7-79	
d, A	I/I_1	d, A	I/I_1
5.2	ms[a]	5.10	50
4.3	w	4.32	10
3.90	ms[a]	3.884	60
3.73	w	3.722	10
3.50	m[a]	3.500	20
3.48	w	3.481	10
3.01	w	3.010	10
2.99	w	2.994	10
2.77	s[a]	2.770	100
2.51	s[a]	2.514	100
2.46	s[a]	2.460	80
2.35	m[a]	2.350	20
2.32	w	2.318	10
2.27	ms[a]	2.271	40
2.25	ms[a]	2.251	30
2.16	w	2.162	10
...	...	2.034	5
1.88	w	1.879	10
1.75	s[a]	1.751	40
1.74	w	1.741	10

[a] High intensity lines chosen for use in search procedure.

FORSTERITE MAGNESIUM ORTHO SILICATE 7-75; 5.10 3.88 2.77 2.51 2.46 2.27 2.25 1.75 FORSTERITE MAGNESIUM ORTHO SILICATE 7-74; 5.10 3.88 2.77 2.51 2.46 2.27 2.25 1.75 FORSTERITE MAGNESIUM ORTHO SILICATE 7-79.

Both forsterite and chrysolite are members of the olivine group of minerals, and both account satisfactorily for all of the lines of the unknown pattern, as shown in Table 4. Because olivine is a mineral which shows considerable variability in composition, it was suspected that patterns of other members of this family might also show reasonable agreement with the unknown pattern, and indeed the following two additional ones occur in group 10: 5.22 3.96 3.55 2.82 2.56 2.50 2.30 1.78 FERROHORTONOLITE IRON MAGNESIUM ORTHO SILICATE 7-163; 5.21 3.94 3.54 2.81 2.55 2.49 2.29 1.77 HORTONOLITE IRON MAGNESIUM ORTHO SILICATE 7-158.

It thus appears quite clear that the unknown is an olivine mineral, and the data also suggest that it is a magnesium rich olivine (assuming the d values are reasonably accurate), although even a rough chemical analysis would be highly desirable in clarifying this point.

The identification of single compounds can generally be accomplished by simple, straightforward procedures like the ones in the above examples; however, the identification of the constituents in even simple mixtures from their diffraction patterns can be a difficult and challenging task. In all cases, except possibly those involving very simple patterns, the possibility that the unknown is a mixture must be presumed, and primary emphasis must be placed on achieving a fully confirmed identification of one of the components, since this will greatly simplify the identification of the remaining components. Therefore, the corresponding card in the File should be checked as soon as a reasonable agreement is found for any entry in the index book. Once one or more components have been identified, careful consideration must be given to designing the search procedure for the remaining constituents. Generally, lines of the unknown pattern which have the highest intensities and largest d values, but

TABLE 5—DATA FOR IDENTIFICATION OF REFLECTION ELECTRON DIFFRACTION PATTERN FROM ETCHED SURFACE OF Fe-Cr-CO-Ni HEAT-RESISTANT ALLOY.

Unknown		TiC, 6-614		TiN, 6-642		$Cr_{23}C_6$, 9-122		$M_{23}C_6$, 11-545	
d, A	I/I_1	d, A	I/I_1	d, A	I/I_1	d, A	I/I_1	d, A	I/I_1
3.7	vw	...	...	...	...	...	...	...	...
3.2	vw	...	...	...	...	3.21	25	...	...
3.1	vw	...	...	...	...	3.07	25	...	...
2.68	vw	...	...	...	...	2.66	50	...	...
2.49	vs	2.50	80	...	...	...	...	...	...
2.46	vw	...	...	2.44	77	2.44	25	...	...
2.39	w	...	...	...	...	2.38	100	2.38	50
2.18	w	...	...	...	...	2.17	100	2.177	60
2.16	s	2.179	100	...	...	...	...	...	...
2.14	w	...	...	2.12	100	...	...	...	...
2.05	m	...	...	...	...	2.05	100	2.053	100
1.88	vw	...	...	...	...	1.88	75	1.881	70
1.81	w	...	...	...	...	1.80	100	1.799	50
...	...	...	...	...	...	1.77	50	...	...
1.67	w	...	...	...	...	1.68	25	1.686	20
1.62	vw	...	...	...	...	...	...	...	...
1.52	m	1.535	50	...	...	...	...	...	...
1.50	wm	...	...	1.49	56	1.49	25	...	...
1.48	vw	...	...	...	...	1.48	25	...	...
1.30	w	1.311	30	...	...	1.38	25	1.330	35
1.29	w	...	...	1.277	26	1.29	75	1.290	30
1.25	wm	1.255	10	...	...	1.25	100	1.254	80
1.23	wm	...	...	1.223	16	1.23	100	1.230	75
...	...	1.086	5	...	...	...	...	...	...

which do not belong to patterns of components already identified, will provide a logical choice for d_s. Allowance must always be made for possible coincidence of lines in several patterns, however, and frequently a line may be chosen as d_s in searching for a component not yet identified when its intensity in the unknown pattern is much greater than would be expected on the basis of the intensity reported for it in the pattern of a constituent already identified. In all cases, however, d values of components already identified must be included in choosing d_{s-1}, d_{s-2}, etc., in steps three to seven of the search procedure outlined above. The procedures initially chosen for the previous two examples would have been appropriate had the unknowns been mixtures; however, the total search procedures were much simplified when it became apparent that only single phases were involved. The following examples illustrate how the procedures become more complicated when mixtures are actually involved.

Example 3—The data in Table 5 are from an electron diffraction pattern obtained by the reflection technique from the surface of a heat-treated specimen of an iron-chromium-cobalt-nickel, heat-resistant alloy, known to contain a few per cent of manganese, silicon, titanium, aluminum, and molybdenum as minor alloying constituents. The specimen had been metallographically polished and then etched so that particles of precipitate phases present in it protruded slightly from the surface so that the diffraction pattern could logically be expected to have been produced by these particles. Of the elements known to be present in the alloy, titanium and chromium are particularly likely to react to form carbides and nitrides. A straightforward procedure to adopt, therefore, would be to chose $d_s = 2.49 \pm 0.03$ and $d_{s-1} = 2.16$ or 2.05, and to search sections 60 (2.51–2.50), 61 (2.49–2.48), and 62 (2.47–2.46) in the regions where d values ranging from 2.00 to 2.20 appear in the second column of the index, looking only for likely carbide and nitride phases initially. This yields the following entries in the index book that show reasonably good agreement with lines of the unknown pattern:

Group 60 (2.51–2.50):

2.50 2.16 2.06 1.70 1.24 1.20
1.16 3.40 IRON CARBIDE 3-400;
2.51 2.18 1.54 1.31 1.26 0.97 0.88
0.83 TITANIUM CARBIDE 6-614.

Checking these possibilities immediately reveals excellent agreement between the data for titanium carbide (TiC) and the five strongest lines of the unknown pattern, as shown in Table 5, strongly suggesting that TiC is the major component contributing to the unknown pattern. The fact that all five d values in the unknown are slightly less than those reported for TiC, however, suggests that the actual phase is probably a carbon-rich carbonitride, since TiC and titanium nitride (TiN) are well known to form a continuous series of solid solutions. The diffraction data for TiN were therefore obtained, by reference to the Davey index, and found to agree satisfactorily with an additional five lines of the unknown, again making allowance for solid-solution effects, as shown in Table 5. Several oxide patterns based on hematite (Fe_2O_3) also showed reasonably good agreement with the unknown; however, these were rejected because alloys of the kind under investigation usually do not show surface oxidation under the etching conditions used, and because the diffraction rings were sharp and spotty, as would be expected for discrete large particles, rather than diffuse and nearly continuous, as would be expected for an oxide film.

In proceeding to identify the remaining components of the unknown, it appears best to choose $d_s = 2.05$, since this line has the highest intensity of the unidentified lines combined with a satisfactorily large d value. Based on this choice, and allowing for considerable uncertainty in the d values, an effective search procedure would be to examine chemical names can be scanned. Using this approach, the entries listed in Table 6 are found which appear to merit further consideration as possible constituents of the unknown. In addition, the following oxides also appeared as possible components, but were rejected for the reason stated above: COBALT-IRON-OXIDE 1-1121 and 3-864, CHROMIC OXIDE

TABLE 6—ENTRIES IN FINK INDEX SHOWING PRELIMINARY AGREEMENT WITH UNKNOWN DATA OF TABLE 5.

				Group 81 (2.09–2.08)					
2.08	1.60	1.37	1.24	1.16	0.00	2.38	2.16	EPSILON IRON CARBIDE	6-670
2.09	1.61	1.37	1.24	1.16	1.14	2.38	2.19	IRON NITRIDE	1-1236
2.08	1.82	1.39	1.36	1.31	1.19	2.29	2.13	TITANIUM SILICIDE	2-1120
2.09	1.82	1.53	1.29	3.28	2.45	2.22	2.12	GAMMA CHROMIUM SILICIDE	12-596
2.09	1.86	1.66	1.47	1.29	2.74	2.54	2.47	MOLYBDENUM MONO CARBIDE	6-546
				Group 82 (2.07–2.06)					
2.07	1.46	1.25	1.20	1.03	0.93	0.85	2.39	CHROMIUM NITRIDE	11-65
2.06	1.70	1.24	1.20	1.16	3.40	2.50	2.16	IRON CARBIDE	3-400
2.07	1.82	1.69	1.24	1.16	1.12	3.37	2.15	IRON CARBIDE	3-410
				Group 83 (2.05–2.04)					
2.04	1.69	1.55	1.24	1.16	1.12	3.37	2.14	IRON CARBIDE	3-411
2.05	1.88	1.80	1.25	1.23	1.09	2.38	2.18	Cr-CO-Ni CARBIDE	11-545
2.05	1.88	1.80	1.29	1.25	1.23	2.38	2.17	CHROMIUM CARBIDE	9-122
				Group 84 (2.03–2.02)					
2.02	1.57	1.32	1.22	1.13	1.11	2.28	2.16	NICKEL CARBIDE	6-697
2.02	1.80	1.20	1.17	1.15	1.15	2.26	2.12	IRON CARBIDE	6-686
2.02	1.48	1.45	1.26	1.22	1.20	2.46	2.10	MANGANESE NITRIDE	1-1158

those portions of groups 81 (2.09–2.08), 82 (2.07–2.06), 83 (2.05–2.04), and 84 (2.03–2.02) where d values ranging from 1.45 to 1.95 appear in the second column, looking again for likely carbides and nitrides of elements known to be present in the alloy. This approach would cover possibilities where any of the 6 lines having d values of 1.88, 1.81, 1.67, 1.62, 1.52, and 1.50 might be listed in the second column of the index, yet the search would not take overly long because of the rapidity with which the 6-532, COBALT OXIDE 9-418, and COBALT-NICKEL-OXIDE 2-1074.

Of the compounds listed in Table 6, the chromium carbide ($Cr_{23}C_6$) and chromium-cobalt-nickel carbide ($M_{23}C_6$) patterns agree best with the unknown. Since these are reasonable precipitate phases to appear in an alloy of this type, the identification was considered completed, and the unknown pattern attributed to two types of precipitate phase, the first being a solid solution titanium carbonitride, and the second being a carbide based on

TABLE 7—DATA ILLUSTRATING THE IDENTIFICATION OF THE COMPONENTS OF BOILER SCALE.

Boiler Scale[a]		Anhydrite, 6-226		Calcite, 5-586		Magnetite, 11-614	
d, A	I/I_1	d, A	I/I_1	d, A	I/I_1	d, A	I/I_1
5.80	1	...	...	...	...	...	...
4.83	3	...	...	...	...	4.85	40
3.87	4	3.87	6	3.86	12	...	...
3.49	25[b]	3.498	100	...	...	...	...
...	...	3.118	3	...	...	...	...
3.03	30[b]	...	...	3.035	100	...	...
2.95	10[b]	...	...	...	...	2.97	70
2.84	15[b]	2.849	33	2.845	3	...	...
...	...	2.797	4	...	...	...	...
2.52	75[b]	2.473	8	2.495	14	2.530	100
2.41	6	...	...	...	...	2.419	10
2.33	8	2.328	22	...	...	...	...
2.27	10[b]	...	...	2.285	18	...	...
2.21	8	2.208	20	...	...	...	...
...	...	2.183	8	...	...	...	...
2.08	30[b]	2.086	9	2.095	18	2.096	70
1.99	1	1.993	6	...	...	...	...
...	...	1.938	4	1.927	5	...	...
1.91	10[b]	...	...	1.913	17	...	...
1.865	12[b]	1.869	15	1.875	17	...	...
...	...	1.852	4	...	...	...	...
1.745	4	1.749	11	...	...	...	...
...	...	1.748	10	...	...	...	...
1.705	6	...	...	...	...	1.712	60
1.640	4	1.648	14	1.626	4	...	...
1.606	20	1.594	3	1.604	8	1.614	85
1.560	1	1.564	5	1.587	5	...	...
...	...	1.525	4	...	...	...	...
1.518	6	1.515	1	1.518	4	...	...
...	...	...	...	1.510	3	...	...
1.480	30[b]	1.490	5	1.473	2	1.483	85
1.439	4	...	...	...	...	...	...
1.420	1	1.424	3	1.422	3	...	...
...	...	1.418	1	...	...	...	...
...	...	1.398	3	...	...	...	...
...	...	1.396	3	...	...	...	...
...	...	1.365	1	1.356	1	...	...
1.320	2	1.319	4	1.339	2	1.327	20
...	...	1.296	2	1.297	2	...	...
1.278	6	1.277	5	1.284	1	1.279	30
1.210	2	etc.		etc.		1.211	20
1.150	2	...		...		...	...
1.120	2	...		...		...	...
1.090	7	...		...		1.092	60
1.047	6	...		...		etc.	
1.010	2	...		...			
etc.							

[a] Data originally presented by Frevel ((**3**)).
[b] High intensity lines selected for use in starting the identification procedure.

$Cr_{23}C_6$ with other metallic elements probably incorporated into it by solid solution.

Example 4—The final example of the use of the Fink index involves the identification of the constituents of a boiler scale from the data of Table 7 which were originally used by Frevel (**3**) to illustrate the Hanawalt index. Reference to Frevel's article will provide a comparison of the use of the Fink and Hanawalt indexing systems.

In addition to the crystallographic data given in Table 7, Frevel also stated that calcium and iron were the principal metallic elements found to be present in the scale by spectrographic analysis, although traces of cobalt, molybdenum, copper, and sodium were also detected. The d values which appear appropriate for use in searching the Fink index are indicated. The most straightforward initial search procedure would be to choose $d_s = 3.49$ and $d_{s-1} = 3.03$, 2.95, or 2.84, and to look for compounds of calcium or iron which are likely components of boiler scale and which otherwise match the unknown pattern. Searching the portions of Groups 25 (3.50–3.54) and 26 (3.45–3.49) where d values ranging from 2.80 to 3.10 are found in the second column, yields the following entries which show reasonable preliminary agreement with the unknown data: 3.48 3.01 2.80 1.85 1.69 1.30 1.14 605 GAMMA CALCIUM SULFATE 2-134; 3.46 3.05 2.85 2.33 2.24 2.09 1.90 4.49 HYDROPHILITE CALCIUM CHLORIDE 1-338; 3.50 2.85 2.33 2.21 1.87 1.75 1.75 1.65 ANHYDRITE CALCIUM SULFATE 6-226. Comparison of data from the corresponding file cards with the unknown data leads to the identification of anhydrite ($CaSO_4$) as one component of the unknown, as shown in Table 7.

With one component identified, it now becomes necessary to decide on a strategy for attempting to identify additional components. Since there are no lines reported for $CaSO_4$ having d values of 3.03, 2.95, 2.52, and 2.27, and since the 2.08 line in the unknown pattern appears much stronger than would be expected if it belonged only to $CaSO_4$, it appears reasonable to use these lines as the basis for the next stage of the search. Choosing $d_s = 3.03 \pm 0.03$ and $d_{s-1} = 2.95 \pm 0.03$, and again looking for likely compounds of calcium or iron, does not yield any likely possibilities. Allowing for possible coincidence of lines, and choosing $d_s = 3.03 \pm 0.03$ and $d_{s-1} = 2.84 \pm 0.03$ is not successful either; however, choosing $d_s = 3.03 \pm 0.03$ and $d_{s-1} = 2.52 \pm 0.03$ yields the entry: 3.04 2.50 2.28 2.10 1.91 1.88 1.60 3.86 CALCITE CALCIUM CARBONATE 5-586, and reference to data card 5-586 confirms the identification of calcite ($CaCO_3$) as a component of the unknown.

Neither of the phases identified so far accounts for the iron found spectrographically, so there is apparently a third phase present in the scale. Careful examination of the data for the unknown and the $CaSO_4$ and $CaCO_3$ patterns in Table 7, assuming the unknown gave a reasonably good pattern and that reasonable care was exercised in obtaining the data from it, suggests that the 2.08, 1.60, and 1.48 lines are relatively more intense in the unknown pattern than would be expected if they were produced only by the anhydrite and calcite phases, while the 2.95, 2.52, 2.41, and 1.70 lines of the unknown are not satisfactorily accounted for by the anhydrite and calcite patterns. These observations suggest a search procedure based on $d_s = 2.95 \pm 0.03$ with $d_{s-1} = 2.52$, 2.41, 2.08, 1.70, 1.60, or 1.48. Searching groups 36 (2.99–2.98), 37 (2.97–2.96), 38 (2.95–2.94), and 39 (2.93–2.92) under this plan, and looking principally for likely compounds of iron, yields as possibilities: 2.99 2.60 2.60 2.56 2.49 1.51 3.01 3.00 CALCIUM-IRON-OXIDE GAMMA 13-395; 2.97 2.53 2.16 1.71 1.61 1.48 1.09 4.85 MAGNETITE IRON OXIDE 11-614. Reference to the data cards quickly confirms the identification of magnetite (Fe_3O_4) as a third constituent of the boiler scale. As shown in Table 7, the combination of $CaSO_4$, $CaCO_3$, and Fe_3O_4 satisfactorily accounts for the spectrographic analysis, and for all of the lines of the diffraction pattern of the unknown, except for the very weak line with $d = 5.80$. Examination of all entries in groups 7 (5.99–5.80) and 8 (5.79–5.60) did not yield any likely compounds to account for this line.

THE MATTHEWS COORDINATE INDEX

The Matthews index is a punched-card, coordinate index based on modern techniques of information retrieval which combines rapid, convenient search procedures with a high degree of flexibility and versatility, yet requires only relatively simple and inexpensive equipment.

General Principles of Coordinate Indexing:

The punched-card coordinate indexing system is both simple and ingenious. The index consists of a number of cards each of which represents one of a corresponding set of properties or descriptors, $D_1, D_2, D_3 \cdots D_i \cdots D_n$, that can be searched for by means of the index. Some of the descriptors which might be used in an index to the File include d values, lattice constants, unit cell type, density, color, refractive index, and chemical composition. All cards of the index have a common set of hole locations which are arranged in parallel rows and columns corresponding to the intersections of two mutually perpendicular sets of parallel coordinate lines. Any hole location is uniquely specified by giving the number of its column (or vertical coordinate line) and row (or horizontal coordinate line). Each item to be covered by the index is assigned a particular hole location which is the same for all cards of the index. In constructing a coordinate index, each item is examined, and a hole is punched at its particular hole location in each index card that represents one of its properties or descriptors. Thus when the index is completed, each card D_i will have holes punched at the locations of all items covered by the index to which the descriptor D_i applies. To use the index to

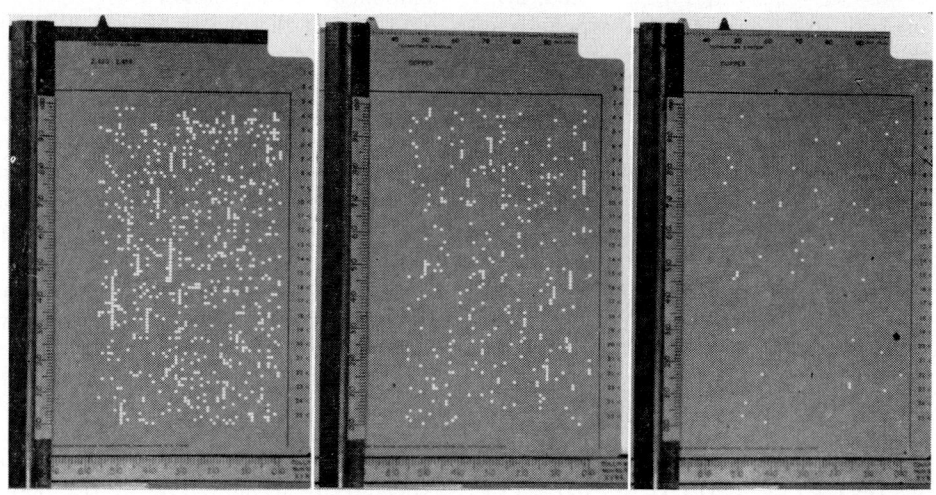

(*left*) Positive d value card Ppl/46.
(*center*) Positive element card Wht/Cu.
(*right*) Cards Ppl/46 and Wht/Cu superimposed.
FIG. 4—Termatrex Cards From Decks A, B, and C of Matthews Coordinate Index.

find all items which possess a given property D_i, it is only necessary to remove the corresponding card D_i from the index, note the row and column number for each hole appearing in the card, and then to refer these numbers to an appropriate listing to determine the identity of the corresponding items. If it is desired to find all items which possess both descriptors D_i and D_j, then the cards D_i and D_j are withdrawn from the index, superimposed, and examined to determine the locations at which they have holes in common. This is conveniently done by observing the superimposed cards against a lighted background so that light shines through the coincident holes. For example, Fig. 4 shows the two cards of the Matthews index for the descriptors, "compound contains the element Cu" and "compound has an intense line in the interval from $d = 2.430$ A to $d = 2.499$ A," and the superposition of these cards to find all compounds listed in the index which "contain the element Cu and have an intense line with the d value in the interval from 2.430 to 2.499 A." It should be noted that the descriptors used in a coordinate index may be either positive or negative in character; that is, descriptor D_j might be, "the element calcium is present," while descriptor D_k might be "the element calcium is not present." In this particular case where D_j and D_k are mutually exclusive, all items containing calcium would be listed on card D_j by punching holes at their corresponding locations, while all remaining items would be listed on card D_k.

Construction of the Matthews Index

The Termatrex Cards:

The Termatrex cards used in the Matthews index are constructed of a highly durable, dimensionally stable, plastic, and are specially shaped to facilitate accurate alignment in the Termatrex card readers and punches.

a	Name of the Substance	Eight Strongest Lines of the Pattern	b
0000	IRON ARSENIDE	2.64 2.59 2.12 2.08 2.02 2.00 1.72 1.69	12-0799
0001	GADOLINIUM OXIDE CHLORIDE	6.68 3.40 2.80 2.55 2.14 1.97 1.71 1.56	12-0798
0002	GADOLINIUM OXIDE	4.42 3.12 2.70 2.55 2.12 1.91 1.63 1.24	12-0797
0003	ZIRCONIUM IODATE	4.18 3.73 2.96 2.78 2.32 2.09 1.67 1.67	12-0796
0004	CESIUM VANADIUM SULFATE 12 HYDRATE	6.23 5.56 4.40 3.59 3.11 2.85 2.78 2.54	12-0795
0005	CESIUM PERCHLORATE	4.30 3.91 3.81 3.63 3.43 3.11 3.02 2.32	12-0794
0006	SILVER OXIDE	3.35 2.73 2.37 1.67 1.43 1.37 1.09 1.06	12-0793
0007	SAMARIUM FLUORIDE	3.56 3.48 3.13 2.01 1.96 1.75 1.69 1.40	12-0792
0008	CERIUM CHLORIDE	6.47 3.73 3.59 2.82 2.59 2.15 2.12 1.41	12-0791
0009	SAMARIUM OXIDE CHLORIDE	6.73 3.43 2.82 2.57 2.16 1.99 1.72 1.57	12-0790
0010	SAMARIUM CHLORIDE	6.39 3.69 3.49 2.76 2.54 2.13 2.09 1.38	12-0789
0011	GADOLINIUM FLUORIDE	3.72 3.65 3.49 3.24 2.97 2.00 1.96 1.90	12-0788
0012	PRASEODYMIUM CHLORIDE	6.46 3.72 3.56 2.81 2.57 2.14 2.11 2.03	12-0787

FIG. 5—Portion of Page From Conversion Table for Matthews Coordinate Index, Relating Coordinate Address Numbers of Holes in Termatrex Cards (a) to the Numbers of Cards in the Data File (b).

Along the top of each card there is a colored border and identification tab, with the latter usually located at a significant position with respect to a scale printed on the border so as to facilitate easy selection of a desired card from the randomly ordered index card deck. Each card has 10,000 hole locations ordinate address number 0973 specifies the hole location which occurs in column 9 and row 73, while the coordinate address number 5623 specifies the hole location occurring in column 56 and row 23. The column and row numbers are simply and easily determined with the Termatrex card reader or with an or-

TABLE 8—d VALUE INTERVALS USED AS DESCRIPTORS FOR CARDS OF DECK A OF THE STANDARD SET OF THE MATTHEWS INDEX.

Card Designator[a]	d Value Interval, A	Card Designator[a]	d Value Interval, A
Ylo/99	10.0 and greater	Blk/32	3.280–3.359
Ylo/92	8.500–9.999	Blk/24	3.210–3.279
Ylo/80	7.500–8.499	Blk/17	3.140–3.209
Ylo/72	6.900–7.499	Blk/10	3.060–3.139
Ylo/66	6.400–6.899	Blk/02	3.000–3.059
Ylo/62	6.000–6.399		
Ylo/58	5.700–5.999	Ppl/95	2.920–2.999
Ylo/54	5.300–5.699	Ppl/88	2.850–2.919
Ylo/51	5.000–5.299	Ppl/81	2.780–2.849
		Ppl/74	2.710–2.779
Grn/90	4.800–4.999	Ppl/67	2.640–2.709
Grn/72	4.650–4.799	Ppl/60	2.570–2.639
Grn/57	4.500–4.649	Ppl/53	2.500–2.569
Grn/45	4.400–4.499	Ppl/46	2.430–2.499
Grn/35	4.300–4.399	Ppl/39	2.360–2.429
Grn/25	4.200–4.299	Ppl/32	2.290–2.359
Grn/15	4.100–4.199	Ppl/25	2.220–2.289
Grn/05	4.000–4.099	Ppl/18	2.150–2.219
		Ppl/11	2.080–2.149
Blk/96	3.920–3.999	Ppl/04	2.000–2.079
Blk/88	3.840–3.919		
Blk/80	3.760–3.839	Org/95	1.900–1.999
Blk/72	3.680–3.759	Org/85	1.800–1.899
Blk/64	3.600–3.679	Org/75	1.700–1.799
Blk/56	3.520–3.599	Org/65	1.600–1.699
Blk/48	3.440–3.519	Org/55	1.500–1.599
Blk/40	3.360–3.439	Org/25	less than 1.5

[a] The letters indicate the color of the tabs and top borders of the cards as follows: Ylo = Yellow, Grn = Green, Blk = Black, Ppl = Purple, Org = Orange. The number following the slant indicates the location of the tab relative to the scale printed on the colored border. The second digit of this number is printed on the tab.

arranged in 100 equally spaced rows and columns. The rows are numbered from 00 to 99, beginning with the bottom row. The columns are numbered from 00 to 99 beginning with the right hand one. Each hole location is specified by a four-digit coordinate address number, the first two digits of which indicate the number of the column in which it occurs, while the last two digits indicate the row in which it occurs. Thus, the co-

dinary light box and T-square, as shown in Fig. 4.

The identity of the card in the Powder Diffraction File which is assigned to any given hole location can be determined by reference to the Conversion Table for the Matthews index. Part of a page from this book is reproduced in Fig. 5. As shown, the right-hand column gives the number of the data card in the data file which is assigned to the hole location whose four-digit coordinate address number is listed in the left-hand column. The name of the compound whose powder diffraction pattern is recorded on the data card is also given, together with the d values of the eight lines of the patterns having the highest relative intensities. These are the same eight d values as are used for the compound in the Fink index. Although no numerical values are given, the relative intensities of most of these lines will be greater than 30. These eight d values are sufficient to eliminate highly unlikely patterns in the preliminary stages of the search, and often lead to a reasonably certain identification of the unknown, thereby greatly reducing the number of patterns to be looked up in the data File itself.

At present, the Matthews index consists of the following five decks of Termatrex index cards:

Deck A...50 positive d-value cards of the Standard Set
Deck B...55 positive element cards of the Standard Set
Deck C...47 supplementary positive d-value cards
Deck D...20 supplementary negative element cards
Deck E...37 experimental negative d-value cards

As indicated, Decks A and B were published together as the Standard Set of cards for the first edition of the Matthews index, and Decks C and D were published separately as supplements to the Standard Set. While Deck E has only recently been prepared for study and evaluation, it will be published as a supplementary part of future editions of the index. As will be evident in the following examples, the supplementary decks contribute so much to the convenience, effectiveness, and efficiency of the index that their use is almost always advantageous when they are available.

Decks A and C:

As indicated in Tables 8 and 9, the

TABLE 9—d VALUE INTERVALS USED AS DESCRIPTIONS FOR CARDS OF SUPPLEMENTARY DECK C OF THE MATTHEWS INDEX.

Card Designator[a]	d Value Interval, A
Ylo/85	8.000–9.199
Ylo/75	7.200–7.999
Ylo/69	6.600–7.199
Ylo/64	6.200–6.599
Ylo/60	5.800–6.199
Ylo/57	5.400–5.799
Ylo/53	5.100–5.399
Grn/99	4.900–5.099
Grn/80	4.720–4.899
Grn/65	4.570–4.719
Grn/50	4.450–4.569
Grn/40	4.350–4.449
Grn/30	4.250–4.349
Grn/20	4.150–4.249
Grn/10	4.050–4.149
Blk/99	3.960–4.049
Blk/92	3.880–3.959
Blk/84	3.800–3.879
Blk/76	3.720–3.799
Blk/68	3.640–3.719
Blk/60	3.560–3.639
Blk/52	3.480–3.559
Blk/44	3.400–3.479
Blk/36	3.320–3.399
Blk/28	3.240–3.319
Blk/21	3.170–3.239
Blk/14	3.100–3.169
Blk/06	3.020–3.099
Ppl/98	2.950–3.019
Ppl/92	2.880–2.949
Ppl/85	2.810–2.899
Ppl/78	2.740–2.809
Ppl/71	2.670–2.739
Ppl/64	2.600–2.669
Ppl/57	2.530–2.599
Ppl/50	2.460–2.529
Ppl/43	2.390–2.459
Ppl/36	2.320–2.389
Ppl/29	2.250–2.319
Ppl/22	2.180–2.249
Ppl/15	2.110–2.179
Ppl/08	2.040–2.109
Org/99	1.950–2.039
Org/90	1.850–1.949
Org/80	1.750–1.849
Org/70	1.650–1.749
Org/60	1.550–1.649

[a] The letters indicate the color of the tabs and top borders of the cards as follows: Ylo = Yellow, Grn = Green, Blk = Black, Ppl = Purple, Org = Orange. The number following the slant indicates the location of the tab relative to the scale printed on the colored border. The second digit of this number is printed on the tab.

descriptors for the index cards of Decks A and C are d-value intervals covering the range above 1.5 A. The catch-all descriptor "less than 1.5 A" is assigned to one card, Org/25, to take care of those few cases (to be discussed later) where it is necessary to utilize d values less than 1.5 A. The color of the upper borders and identification tabs of the cards indicates the general portion of the d-value range in which their descriptors fall; the locations of the identification tabs with respect to a numerical scale printed on the top borders of the cards, combined with the color coding, give approximately the midrange of the d-value interval assigned to the cards and also form the card designators given in the tables. These relationships are indicated in Table 10. For intervals below 5.000 A the color code indicates the integers for the midrange values, while the tab locations and designator numbers both indicate the first two decimal places for the midrange values. Thus the descriptor for card Grn/35 is the d-value interval from 4.300 to 4.399 A whose midrange is approximately 4.35 A, while the descriptor for card Ppl/57 is the interval from 2.530 to 2.599 A whose midrange is approximately 2.57 A. For intervals above 5.000 A, all index cards have yellow tabs, and the integers and first decimal places of their descriptor midranges are indicated by the card designator numbers and the tab locations. Thus, the descriptor for card Ylo/64 is the d-value interval from 6.200 to 6.599 A whose mid-range is approximately 6.4 A. This system allows the cards for Decks A and C to be intermingled in a random order in the Termatrex card holder, and yet permits any desired card to be selected from the combined decks quickly and accurately.

The descriptor d-value intervals for the d-value range above 1.5 A vary in size in accordance with the distribution of strong lines in the diffraction patterns, so that roughly the same number of holes appear in each card of the index. In general, the intervals are from 2 to 4

TABLE 10—RELATIONSHIPS BETWEEN COLOR CODING, d VALUE INTERVAL, INTERVAL MIDRANGE, AND CARD DESIGNATORS FOR CARDS OF DECKS A AND C OF MATTHEWS INDEX.

Color	d Value Interval, A	Card Designator	Interval Midrange, A	Tab Location
Yellow	5.000 and greater	Ylo/XY	XY	XY
Green	4.000 to 4.999	Grn/XY	4.XY	XY
Black	3.000 to 3.999	Blk/XY	3.XY	XY
Purple	2.000 to 2.999	Ppl/XY	2.XY	XY
Orange	1.500 to 1.999	Org/XY	1.XY	XY
	less than 1.5	Org/25	...	25

per cent of the interval mid-range value. The intervals for the cards in Deck C overlap those for Deck A by one-half so that it is possible to choose an interval that will extend far enough on either side of any d value in an unknown pattern to allow for normal experimental variability and uncertainty. Intervals narrower than those provided by the individual cards can be obtained by using two cards with overlapping intervals. These relationships are illustrated for the two cards Org/70 and Org/75.

d-value interval for Org/70.......1.650–1.749
d-value interval for Org/75.......1.700–1.799
d-value interval for Org/70 + Org/75............................1.700–1.749

In general, the listing of the patterns contained in the File on the index cards of Decks A and C was based on the three, four, or five strongest lines of the patterns having d values greater than 1.5 A. The instructions used in selecting these d values are outlined as follows:

1. Use all d values when there are less than five in the pattern.

2. Use all d values with $I/I_1 = 100$, even if there are more than five of these.

3. Otherwise select the five d values above 1.5 A which have the highest relative intensities, and which are not within 1 per cent of each other in d values.

4. When there are not five d values above 1.5 A, select from all d values given those five which have the highest relative intensities and which are not within 1% of each other in d values. If three, four, or five of these d values have relative intensities greater than 30, use only these. If only 1 or 2 of these d values have relative intensities greater than 30, use all

TABLE 11—EXAMPLE APPLICATIONS OF PROCEDURES FOR LISTING PATTERNS ON INDEX CARDS OF DECKS A AND C OF MATTHEWS INDEX.

Compound	SiI_4		$16\ Al(OH,F)_3 6\cdot H_2O$		β-AgZn	
Data card no.	04-0487		04-0196		08-222	
	d, A	I/I_1	d, A	I/I_1	d, A	I/I_1
Powder data	3.47	100[a]	5.70	100[a]	2.230	100[a]
	3.00	60[a]	2.98	48[a]	1.575	50[a]
	2.13	100[a]	2.84	22[a]	1.410	30
	1.80	100[a]	2.46	6	1.289	100[a]
	1.73	20	2.26	5	1.116	50
	1.38	60	2.01	...	1.050	10
	1.34	50	1.89	20[a]	0.998	100[a]
	1.23	50	1.73	24[a]	0.951	10
	1.16	50	1.66	7	0.911	100[a]
	1.00	30	1.55	6		
Hole location	4266		4315		2697	
Index cards punched in Decks A, B, and C	Blk/48		Ylo/57		Ppl/25	
	Blk/44		Ylo/58		Ppl/22	
	Blk/02		Ppl/98		Org/60	
	Ppl/98		Ppl/95		Org/55	
	Ppl/11		Ppl/85		Org/25	
	Ppl/15		Ppl/81		Wht/Ag	
	Org/85		Org/90		Wht/Zn	
	Org/80		Org/85		Snd/O5	
	Wht/Si		Org/75			
	Wht/I		Org/70			
			Wht/Al			
			Wht/O			
			Wht/F			
			Wht/H			
			Snd/10			
Index cards not punched in Decks D and E; all others punched	Blu/15		Blu/50		Blu/23	
	Blu/21		Blu/29		Blu/15	
	Blu/30		Blu/28		Blu/Au	
	Blu/34		Blu/15		Blu/Zn	
	Blu/Si		Blu/B			
	Blu/F		Blu/O			
			Blu/H			
			Blu/F			

[a] d values selected for indexing according to procedures outlined on page 78.

five of the chosen d values, irrespective of their intensities.

It will be noted that these instructions include special provisions for patterns which have less than three strong lines with d values greater than 1.5 A, and for

assigned to the particular compound in all cards whose descriptor d-value interval included one of the three, four, or five d values selected by the above procedures. In the examples of Table 11, the hole location assigned to each compound

TABLE 12—ELEMENTS AND GROUPS OF ELEMENTS USED AS DESCRIPTORS FOR THE POSITIVE ELEMENT CARDS IN DECK B OF THE STANDARD SET OF THE MATTHEWS INDEX.

Card Designator[a]	Elements Present	Card Designator[a]	Elements Present
What/Ag	silver	Wht/Li	lithium
Wht/Al	aluminum	Wht/Mg	magnesium
What/As	arsenic	Wht/Mn	manganese
Wht/Au	gold, iridium, osmium, rhenium, rhodium, euthenium, technetium	Wht/Mo	molybdenum
		Wht/N	nitrogen
		Wht/Na	sodium
Wht/B	boron	Wht/Nb	niobium, tantalum
Wht/Ba	barium	Wht/Ni	nickel
Wht/Be	beryllium	Wht/O	oxygen
Wht/Bi	bismuth	Wht/P	phosphorus
Wht/Br	bromine	Wht/Pb	lead
Wht/C	carbon	Wht/Po	polonium, selenium, tellurium
Wht/Ca	calcium	Wht/RE	rare earths: Ce, Dy, Er, Eu, Gd, Ho, La, Lu, Nd, Pm, Pr, Sm, Tb, Tm, Yb
Wht/Cd	cadmium		
Wht/Cl	chlorine		
Wht/Co	cobalt	Wht/S	sulfur
Wht/Cr	chromium	Wht/Sb	antimony
Wht/Cs	cesium, rubidium	Wht/Sc	scandium, yttrium
Wht/Cu	copper	Wht/Si	silicon
Wht/F	fluorine	Wht/Sn	tin
Wht/Fe	iron	Wht/Sr	strontium
Wht/Ga	gallium, indium, thallium	Wht/Ti	titanium
Wht/Ge	germanium	Wht/V	vanadium
Wht/H	hydrogen	Wht/W	tungsten
Wht/Hf	hafnium, zirconium	Wht/Zn	zinc
Wht/Hg	mercury	Wht(+)	misc.: Ac, At, Am, Cm, Fr, Np, Pa, Pd, Pu, Ra, Th, U[b]
Wht/I	iodine		
Wht/K	potassium		
Snd/00			compound is a mineral
Snd/05			compound is an alloy
Snd/10			compound is a hydrate
blue transparent			compound contains hydrogen
yellow transparent			compound contains oxygen

[a] The letters Wht and Snd indicate that the tab and top borders of the cards are White and Sand Colored respectively. The chemical element symbols and the second digit of the numbers following the slant are printed on the tabs.

[b] The inert gas elements and elements beyond Cm are not included.

patterns in which the d value of the strongest line, $I/I_1 = 100$, is less than 1.5 A. Examples showing the application of these instructions to patterns for three typical compounds are given in Table 11. In listing a compound on the index cards, a hole was punched at the hole location

is listed, together with the cards which were punched at these locations in constructing the index.

Decks B and D:

The descriptors and designators for the index cards of Decks B and D are listed in Tables 12, 13, and 14. Cards of Deck B allow information about elements present in the unknown to be incorporated into the search procedure, while cards of Deck D allow information concerning elements not present to be used also. It will be noted that more than one chemical element is included in the descriptors for several cards of Deck C, and for most of the cards of Deck D. These multiple assignments must be taken into account in using these cards. Table 11 indicates the cards in these decks which were punched for the three compounds considered.

Experimental Deck E:

The descriptors and designators for the index cards of experimental Deck E are listed in Table 15. In preparing these cards, the preceding rules were applied to the powder diffraction data for each compound in the data File, and in each case the index card D_i was punched at the appropriate hole location when these rules did not lead to the selection of a d value in the descriptor d-value interval for that card. Table 11 indicates the cards that were punched for the three compounds considered there. Thus the use of the cards of Deck E permits information about the absence of strong lines in various d-value intervals of the unknown pattern to be incorporated directly into the search procedure.

USE OF THE MATTHEWS COORDINATE INDEX

The procedure to be followed in using the Matthews index for the identification of an unknown material from its powder diffraction pattern will involve the following general steps:

1. Note all information available concerning the chemical composition of the unknown and select the appropriate cards from Decks B and D to cover this information.

2. Remove from Deck E all cards whose d-value intervals include any of the lines of the unknown pattern, and reserve the remainder of the deck for use in the search procedure.

3. Select from Decks A and C positive

TABLE 13—ELEMENTS AND GROUPS OF ELEMENTS USED AS DESCRIPTORS FOR THE NEGATIVE ELEMENT CARDS IN SUPPLEMENTARY DECK D OF THE MATTHEWS INDEX.

Card Designator[a]	Elements Not Present
Blu/Au	gold, technetium, platinum, iridium, ruthenium, rhodium, osmium, palladium, copper, silver, rhenium
Blu/B	boron, aluminum, gallium, indium, thallium
Blu/Be	beryllium, magnesium
Blu/C	carbon
Blu/Ca	calcium, barium, strontium
Blu/Cr	chromium, molybdenum, tungsten
Blu/F	fluorine, bromine, chlorine, iodine
Blu/H	hydrogen
Blu/Li	lithium, cesium, rubidium, potassium, sodium
Blu/+	misc: Ac, At, Am, Cm, Fr, Np, Po, Pd, Pu, Ra, Th, U
Blu/Mn	manganese, cobalt, iron, nickel
Blu/N	nitrogen, arsenic, bismuth, phosphorus, antimony
Blu/O	oxygen
Blu/RE	rare earths, scandium, yttrium
Blu/S	sulfur, polonium, selenium, tellurium
Blu/Si	silicon, germanium
Blu/Sn	tin, lead
Blu/Ti	titanium, hafnium, zirconium
Blu/V	vanadium, niobium, tantalum
Blu/Zn	zinc, cadmium, mercury

[a] The letters Blu indicate that the tabs and upper borders of the cards are colored blue. The symbols for the chemical element following the slant are printed on the tabs.

d-value cards having d-value intervals which include all of the high-intensity lines (that is, $I/I_1 \geqq 30$) of the unknown pattern. Note that, because of the coding procedure used, lines having d values greater than 1.5 A will be most useful.

4. Superimpose these index cards in appropriate combinations on the card reader and determine the coordinate

address number of coincident holes for each combination used.

5. Check the coordinate address entry in the book containing the Conversion Table for the Matthews Coordinate Index to determine the identity of the compound assigned to each of the coincident holes, and compare the eight d values listed in the Conversion Table with the d values of the unknown pattern to effect a preliminary identification.

6. Compare the unknown pattern with the data on the card in the File for all compounds showing reasonable agreement with the data in the Conversion Table to effect the final identification.

A simple example of this general procedure is provided by the following identification of an electron diffraction pattern obtained from the surface of a copper alloy.

TABLE 14—ASSIGNMENT OF INDIVIDUAL ELEMENTS TO VARIOUS NEGATIVE ELEMENT CARDS OF SUPPLEMENTARY DECK D OF THE MATTHEWS INDEX.

Element Not Present	Card Designator[a]	Element Not Present	Card Designator[a]
Aluminum, Al	Blu/B	Molybdenum, Mo	Blu/Cr
Antimony, Sb	Blu/N	Nickel, Ni	Blu/Mn
Arsenic, As	Blu/N	Niobium, Nb	Blu/V
Barium, Ba	Blu/Ca	Nitrogen, N	Blu/N
Beryllium, Be	Blu/Be	Osmium, Os	Blu/Au
Bismuth, Bi	Blu/N	Oxygen, O	Blu/O
Boron, B	Blu/B	Palladium, Pd	Blu/Au
Bromine, Br	Blu/F	Phosphorus, P	Blu/N
Cadmium, Cd	Blu/Zn	Platinium, Pt	Blu/Au
Calcium, Ca	Blu/Ca	Polonium, Po	Blu/S
Carbon, C	Blu/C	Potassium, K	Blu/Li
Cesium, Cs	Blu/Li	Rare Earths, RE	Blu/RE
Chlorine, Cl	Blu/F	Rhenium, Re	Blu/Au
Chromium, Cr	Blu/Cr	Rhodium, Rh	Blu/Au
Cobalt, Co	Blu/Mn	Rubidium, Rb	Blu/Li
Copper, Cu	Blu/Au	Ruthenium, Ru	Blu/Au
Fluorine, F	Blu/F	Scandium, Sc	Blu/RE
Gallium, Ga	Blu/B	Selenium, Se	Blu/S
Germanium, Ge	Blu/Si	Silicon, Si	Blu/Si
Gold, Au	Blu/Au	Silver, Ag	Blu/Au
Hafnium, Hf	Blu/Ti	Sodium, Na	Blu/Li
Hydrogen, H	Blu/H	Strontium, Sr	Blu/Ca
Indium, In	Blu/B	Sulfur, S	Blu/S
Iodine, I	Blu/F	Tantalum, Ta	Blu/V
Iridium, Ir	Blu/Au	Technetium, Tc	Blu/Au
Iron, Fe	Blu/Mn	Tellurium, Te	Blu/S
Lead, Pb	Blu/Sn	Thallium, Tl	Blu/B
Lithium, Li	Blu/Li	Tin, Sn	Blu/Sn
Magnesium, Mg	Blu/Be	Titanium, Ti	Blu/Ti
Manganese, Mn	Blu/Mn	Tungsten, W	Blu/Cr
Mercury, Hg	Blu/Zn	Vanadium, V	Blu/V
Misc.,		Yttrium, Y	Blu/RE
Ac, At, Am, Cm, Pr, Np, Pa,		Zinc, Zn	Blu/Zn
Pd, Pu, Ra, Th, U	Blu/+	Zirconium, Zr	Blu/Ti

[a] The letters Blu indicate that the tabs and top borders of the cards are blue. The symbols for the chemical elements following the slant are printed on the tabs.

Since it is highly probable that the pattern was produced by a copper compound, the card Wht/Cu (copper present) was selected first. Using this card with the cards Ppl/46 and Ppl/15, corresponding to the two strongest lines of the pattern gave the eight coincident holes listed in Table 17.

TABLE 15—d VALUE INTERVALS USED AS DESCRIPTORS FOR NEGATIVE d VALUE CARDS OF EXPERIMENTAL DECK E OF THE MATTHEWS INDEX.

Card Designator[a]	d Value Interval, A
Blu/15	1.400–1.999
Blu/20	2.000–2.099
Blu/21	2.100–2.199
Blu/22	2.200–2.299
Blu/23	2.300–2.399
Blu/24	2.400–2.499
Blu/25	2.500–2.599
Blu/26	2.600–2.699
Blu/27	2.700–2.799
Blu/28	2.800–2.899
Blu/29	2.900–2.999
Blu/30	3.000–3.099
Blu/31	3.100–3.199
Blu/32	3.200–3.299
Blu/33	3.300–3.399
Blu/34	3.400–3.499
Blu/35	3.500–3.599
Blu/36	3.600–3.699
Blu/37	3.700–3.799
Blu/38	3.800–3.899
Blu/39	3.900–3.999
Blu/40	4.000–4.099
Blu/41	4.100–4.199
Blu/42	4.200–4.299
Blu/43	4.300–4.399
Blu/44	4.400–4.499
Blu/45	4.500–4.599
Blu/46	4.600–4.699
Blu/47	4.700–4.799
Blu/48	4.800–4.899
Blu/49	4.900–4.999
Blu/50	5.000–5.999
Blu/60	6.000–6.999
Blu/70	7.000–7.999
Blu/80	8.000–8.999
Blu/90	9.000–9.999
Blu/99	10.000 and over

[a] The letters Blu indicate that the tabs and upper borders of the cards are colored blue. The numbers following the slant indicate the location of the tab relative to the scale, printed on the colored borders. The second digit of this number is printed on the tab. These descriptors apply to the experimental deck, and may be changed for the cards ultimately published.

Comparing the d values listed in the Conversion Table with those of the unknown pattern strongly suggested that the unknown was copper oxide. This identification was confirmed by reference to Card 05-0667 in the data File. It is also interesting to note that in this case, the use of the card Org/55 in addition to the above three would further reduce the number of coincident holes to the one at location 3870, which corresponds to copper oxide.

In most cases of practical interest search procedures will not be as straightforward and simple as this example, nor

TABLE 16—DATA FOR ELECTRON DIFFRACTION PATTERN FROM COPPER ALLOY AND CARDS CHOSEN FROM MATTHEWS INDEX FOR IDENTIFICATION PROCEDURE.

Unknown		Index Cards Selected for Use in Search Procedure	
d, A	I/I_1		
3.04	vvw		
2.47	s	Ppl/46	(2.430–2.499)
2.12	m	Ppl/15	(2.110–2.179)
1.75	vvw		
1.52	m	Org/55	(1.500–1.599)
1.30	m		
1.05	vvw		
		Wht/Cu	(Copper Present)

TABLE 17—COMPOUNDS AND FILE CARDS CORRESPONDING TO COINCIDENT HOLES OBTAINED IN IDENTIFYING UNKNOWN OF PREVIOUS TABLE USING MATTHEWS INDEX.

Hole Location	Compound	File Card
5576	AMMONIUM FLUO-CUPRATE DI HYDRATE	01-0304
4807	COPPER PHOSPHIDE	02-1263
3870	COPPER OXIDE CUPRITE	05-0667
3158	COPPER PALLADIUM	07-0138
2265	COPPER ALUMINATE	09-0185
1527	COPPER SILVER SELENIDE EUCAIRITE	10-0451
0296	COPPER YTTRIUM	12-0503
0016	COPPER HYDROXIDE SULFATE LANGITE	12-0783

will they lead to the single correct answer. In cases where mixtures are likely, considerable ingenuity is often required to effect an identification of all phases present. Because of the great versatility in search procedures allowed by the Matthews index and the considerable range of unknowns that may be encountered by users of the index, it does not appear possible to recommend a single search procedure that will be effective in all cases. However, certain approaches to search procedures can be suggested on the basis of the construction and characteristics of the index which may prove generally useful in developing search procedures for specific cases. For this purpose, it is useful to consider the identification procedure as consisting of the following three consecutive stages: (1) a preliminary stage involving the use of the Termatrex cards, (2) an intermediate stage involving a semifinal identification of the unknown from information contained in the Conversion Book, and (3) the terminal stage involving the final identification from data listed on the cards in the File.

In general, the identification of single compounds can be accomplished by simple, straightforward procedures like the one in the above example; however, the identification of the components of even simple binary mixtures frequently presents a challenging and difficult problem. Since the unknown must be presumed to be a mixture in most cases, particularly careful attention must be given to the choice of index cards for use in the preliminary stage of the search, for the use of a single incorrect card usually will eliminate the correct solution. The problem, then, is to choose cards whose descriptors apply to one of the components without eliminating that component by also choosing a card whose descriptor applies exclusively to some other component. Because of these complications, and of the possibility that more than one of a group of related compounds may fit the data, it is neither practical nor desirable to attempt to eliminate all but a single choice in the preliminary stage of the search. Instead, the index cards can serve most effectively to provide a rapid and convenient means for reducing the possible choices to some convenient number which can all be considered in detail in the subsequent stages of the identification procedure. Typically it appears that

(*left*) Card Blu/RE from Deck D.
(*right*) Card Blu/39 from Deck E.
FIG. 6—Termatrex Cards From Decks D and E of the Mathews Coordinate Index.

this number will be in the range from 20 to 50. Attempts to achieve further reduction involve such high probabilities of eliminating the correct answer as to offer little potential for total saving in time and effort.

There are several effective means for achieving a reduction in the total number of possible choices with a minimum risk of eliminating the correct ones. Probably the best method is to employ as many as possible of the negative element and negative d-value cards of Decks D and E. As shown in Fig. 6, each of these cards eliminates only a few possibilities; however, with even a little knowledge about the characteristics of the unknown, enough of these negative cards can be chosen to eliminate collectively from 50 to 98 per cent of the total compounds covered by the File with virtually no risk of involving the compounds of the unknown. To do this requires careful attention to the cards selected to avoid unintentionally removing desired compounds. In particular, nearly all of the negative element cards represent several elements, all of which must be considered carefully before any card is selected for use. Otherwise, for example, the card Blu/S might be chosen particularly to eliminate unlikely compounds of polonium, selenium, and tellurium without recognizing that it also eliminates, unintentionally, all sulfates. Likewise, it is best to make a very generous allowance for experimental uncertainties in choosing negative d-value cards, and for a truly conservative approach, no card should be used which covers any line in the unknown pattern, even those having the weakest intensities. Even with such an approach, however, it will usually be possible to select 10 or more cards from each of Decks D and E in a typical case, and thereby to eliminate all but a few hundred compounds from consideration.

After selection of the negative d value and negative element cards, the next step should probably be to select appropriate positive element and positive d-value cards from Decks A, B, and C for use in the search. These cards are relatively much more selective than the negative element and negative d-value cards since each eliminates a large number of possible choices. Compare the small number of holes in the cards of Fig. 4 with the large number in those of Fig. 6. In addition, it is usually more difficult to choose several of these cards which all apply to the same component of the unknown. It therefore appears desirable to use no more of the positive element and positive d-value cards than is absolutely necessary to reduce the number of choices to a convenient few. Whenever possible, some advantage can be gained by choosing as the first positive d-value card used, one whose d-value interval includes two of the strong lines of the unknown pattern, since this will not involve the risk of eliminating a component in the event that the two lines belong to patterns of different components of the mixture. Of course both lines could belong to the same component, in which case no gain will result. A significant reduction in possible choices can also be achieved by using two cards with overlapping d-value intervals to produce narrow descriptor intervals for one or more of the d-values of the unknown pattern; however, this involves the risk of eliminating the correct solution if the data for the unknown or the reference pattern are not sufficiently accurate.

In all cases where there is a chance that the unknown is a mixture, primary emphasis should be placed on achieving an identification of one of the components of the mixture, since this will greatly simplify the identification of the remaining components. Therefore, the intermediate and final stages of the identification should be carried through fully and carefully as soon as a satisfactory reduction in the number of choices is achieved using the index cards. Even if this does not lead to a final identification of one component in any given case, the negative information gained can often be employed to advantage in redesigning

TABLE 18—DIFFRACTION DATA FOR UNKNOWN BOILER SCALE AND CARDS CHOSEN FROM MATTHEWS INDEX FOR USE IN THE IDENTIFICATION PROCEDURE.

Unknown		Cards Chosen From Decks A and C[a]	Cards Not Used From Deck E[b]
d, Å	I/I_1		
5.80	1	...	Blu/50
4.83	3	...	Blu/47, 48
3.87	4	...	Blu/38, 39
3.49	25[a]	Blk/48	Blu/34, 35
3.03	30[a]	Blk/02	Blu/30
2.95	10[a]	Ppl/95	Blu/29
2.84	15[a]	Ppl/85	Blu/28
2.52	75[a]	Ppl/53	Blu/24, 25
2.41	6	...	Blu/23, 24
2.33	8	...	Blu/22, 23
2.27	10[a]	Ppl/29	Blu/22
2.21	8	...	Blu/21, 22
2.08	30[a]	Ppl/08	Blu/20, 21
1.99	1	...	Blu/15
1.91	10	Cards Chosen from Decks B and D	
1.865	12		
1.745	4	Wht/Ca and Fe	
1.705	6	Blu/Au, B, Be, Cr, Li, RE,	
1.640	4	Sn, Ti, V, Zn, +	
1.606	20[a]	Org/70	...
1.560	1	...	...
1.518	6	...	...
1.480	30	...	...
1.439	4	...	...
etc.	etc.		

[a] High intensity lines selected for use in search procedure.
[b] All others used.

the search procedure. Once one or more components of a mixture has been identified, the search procedure for the identification of additional components should be based on the lines of the unknown pattern having the highest intensities and largest d values which do not belong to the patterns of the components already identified, but making due allowance for possible coincidences of lines in the several patterns, particularly in carrying out the intermediate and final stages of the subsequent search. The two following examples illustrate these search procedures.

Example 5—This example will show the way in which the Matthews index might be used to identify the components of the boiler scale, which was the subject of Example 4 for the Fink index. Table 18 shows the data for the unknown pattern plus the cards selected from the various decks of the Matthews index for the search procedure. In the case of Deck E, all cards were removed whose d-value intervals included any line of the unknown pattern, and the remainder of the cards of the deck were combined with the eleven cards chosen from Deck D in the first stage of the search, leaving about 100 coincident holes. Because the spectrographic analysis indicated calcium and iron as the principal metallic elements present, the positive card for calcium (Wht/Ca) was next added leaving 19 coincident holes. Since this is a conveniently small number, it was decided to check all 19. The eighth and ninth holes checked were those for Anhydrite ($CaSO_4$, Hole No. 3724, Card No. 6-0226) and Calcite ($CaCO_3$, Hole No. 3924, Card No. 5-0586), both of which show sufficient agreement with the unknown pattern to be accepted as components of the boiler scale, as shown in Table 7. It would have been possible, of course, to reduce the number of coincident holes below 19 by adding some of the positive d-value cards and this would eventually have led to the identification of either calcite or anhydrite; however, it might not have yielded both unless an unusual stroke of luck prompted continuation of the search after the first of these compounds had been identified. Furthermore, the total time saved would probably not have been very great, since most of the 17 unacceptable possibilities can be rejected on the basis of the eight d values given in the Conversion Table, which can be located and checked quite easily without reference to the cards in the File.

In proceeding with the analysis, the positive calcium card, Wht/Ca, was replaced by the positive iron, Wht/Fe, and the negative calcium, Blu/Ca, cards to look for a component containing iron, since this was the other principal metallic element in the boiler scale. This left about 50 coincident holes, and if necessary all could have been checked with relative ease; however, examination of the data of Table 7 shows that neither $CaCO_3$ nor the $CaSO_4$ phases accounts satisfactorily for the lines with d values of 2.52, 2.08, and 1.48 A, which are among the strongest lines of the unknown pattern. Therefore cards Ppl/53 (2.500–2.564) and Ppl/08 (2.040–2.109) were added leaving only two coincident holes at locations 0881 and 4512. The first of these was for the compound magnetite, Fe_3O_4, 11-0614, which accounts for the remaining lines of the unknown pattern, as shown in Table 7.

This approach using primarily the negative element and negative d-value cards, with as few positive element and positive d-value cards as possible in the initial stage of the search, appears to be the most widely applicable procedure developed for use of the Matthews index to date. Based on the limited experience presently available, it appears that this approach will almost always lead to the identification of the major components of the unknown, and that it will also permit identification of minor components in a majority of cases, provided that sufficient allowance is made for experimental uncertainties in selecting the negative d-value cards (that is, as in removing both cards Blu/38 and Blu/39 to avoid all possibilities of eliminating the 3.87 line of the pattern in Table 18), and that comparable care is exercised in choosing the negative element cards to avoid unintentionally eliminating likely elements.

This procedure would be expected to fail principally in the case where a minor component of a mixture is present in such small quantity that one or more of the lines coded by the instructions outlined previously do not appear in the unknown pattern. This situation would be most likely to occur for compounds such as aluminum fluohydroxide hydrate, 16 Al $(OH,F)_3 \cdot 6$ H_2O, of Table 9

TABLE 19—DISTRIBUTION OF INTENSITIES OF WEAKEST CODED LINES FOR 100 RANDOMLY CHOSEN PATTERNS, AND ESTIMATED MINIMUM PERCENTAGE OF COMPOUND REQUIRED FOR DETECTION OF THESE LINES.

I/I_1 of Weakest Line Coded	Approximate Percentage of Patterns	Minimum[a] Percentage Detectable
70–100	20	2–3
40–70	35	3–5
30–40	35	5–7
20–30	5	7–10
10–20	5	10–20

[a] Estimated values assuming good patterns with no abnormal scattering or absorption and detectability at $1/50$ the intensity of the strongest line of the pattern.

whose patterns have only one or two really intense lines, so that three or four lines of relatively low intensity are coded. This should not cause major difficulty, however, for as shown in Table 19, only about 10 per cent of the patterns in the File have coded lines whose intensities are less than 30, and even the weakest coded lines of these patterns should be detectable in the pattern from unknown mixtures if the corresponding compounds are present in the mixtures to the extent of from 10 to 20 per cent. If present in smaller amounts, these compounds could be eliminated from consideration by the negative d-value cards, provided that the missing weak coded lines are not close

TABLE 20—DATA ILLUSTRATING THE IDENTIFICATION OF COMPONENTS OF PAINT PIGMENT.

Unknown		Pb_3O_4		Fe_2O_3		$ZnCrO_4$		ZnO		PbO	
d, A	I/I_1	d, A	I/I_1	d, A	I/I_1	d, A	I/I_1	d, A	I/I_1	d, A	I/I_1
9.6	15	...	...	...	...	9.5	38	...	...	...	...
6.4	5	6.2	11	...	...	...	...	...	...	5.89	6
4.7	15	...	...	...	...	4.67	38	...	...	...	...
3.67	30	3.66	3	3.68	70	...	...	...	...	...	...
3.39	100	3.38	100	...	...	...	...	...	...	...	...
3.29	5	3.28	7	...	...	...	...	...	...	...	...
3.10	20	3.11	19	...	...	3.10	23	...	...	...	...
3.08	10	...	...	...	...	...	...	...	...	3.07	100
2.94	2	...	...	...	...	...	...	...	...	2.94	31
2.91	40	2.90	48	...	...	...	...	...	...	...	...
2.83	15	...	...	...	...	...	...	2.82	71	...	...
2.77	40	2.78	45	...	...	...	...	...	...	...	...
2.75	2	...	...	...	...	...	...	...	...	2.74	28
2.69	100	...	...	2.69	100	2.67	100	...	...	...	...
2.62	30	2.63	30	...	...	...	...	...	...	...	...
2.60	10	...	...	...	...	...	...	2.60	56	...	...
2.51	35	...	...	2.51	80	2.53	3	...	...	...	...
2.48	20	...	...	...	...	...	...	2.48	100	2.49	1
2.44	5	2.44	2	...	...	2.44	15	...	...	...	...
...	...	...	...	...	...	2.34	3	...	...	2.38	20
2.26	5	2.29	4	...	...	...	...	...	...	2.27	1
...	...	2.26	8	...	...	...	...	...	...	...	...
2.23	5	...	...	...	...	2.24	13	...	...	...	...
2.20	20	2.20	1	2.20	70	...	...	...	...	2.20	1
2.14	15	...	...	...	...	2.14	43	...	...	...	...
...	...	2.07	1	2.07	10	...	...	...	...	...	...
2.04	10	2.03	12	...	...	...	...	...	...	...	...
1.98	10	1.97	12	...	...	1.95	5	...	...	...	...
1.90	20	1.90	22	...	...	...	...	1.91	29	1.96	2
1.84	25	...	...	1.84	70	1.85	5	...	...	1.85	14
1.82	20	1.83	21	...	...	...	...	...	...	1.79	14
1.76	30	1.76	30	...	...	1.75	3	...	...	1.72	15
1.69	40	...	...	1.69	80	...	...	...	...	...	...
1.63	15	...	...	1.63	10	...	...	1.63	40	1.64	13
1.60	3	...	...	...	...	1.66	13	...	...	...	...
1.59	20	1.58	12	1.59	40	...	...	...	...	...	...
1.57	10	...	...	...	...	1.57	25	...	...	...	...
1.55	5	...	...	...	...	...	...	...	...	...	...
1.48	30	...	...	1.48	70	...	...	1.48	35	...	...
1.47	10	...	...	...	...	1.47	20	...	...	...	...
1.45	25	...	...	1.45	80	...	...	...	...	...	...
1.41	10	...	...	...	...	1.40	5	1.41	6	...	...
1.38	5	...	...	...	...	...	...	1.38	28	...	...
1.36	15	...	...	1.35	20	1.35	5	1.36	14	...	...
1.31	20	...	...	1.31	40	...	...	1.30	3	...	...
1.26	26	...	...	1.26	30	...	...	...	...	...	...
1.22	2	...	...	1.22	10	...	...	...	...	...	...
1.21	2	...	...	1.21	10	...	...	1.09	10	...	...
1.18	10	...	...	1.18	30	...	...	1.04	10	...	...
1.16	15	...	...	1.16	30	...	...	...	...	...	...
1.14	35	...	...	1.14	40	...	...	...	...	...	...
1.10	30	...	...	1.10	40	...	...	...	...	...	...
1.05	40	...	...	1.05	50	...	...	...	...	...	...
1.04	5	...	...	1.04	10	...	...	...	...	...	...

enough to any of the lines actually present in the patterns to be included in the interval of the negative d-value cards removed to avoid eliminating the latter lines. Situations involving such a concurrence of unfortunate circumstances will undoubtedly be encountered, but they should be relatively rare; on the other hand, about 90 per cent of the compounds covered by the File have all coded lines with intensities greater than 30, and these should not be excluded from consideration by the above recommended search procedure, even when they constitute as little as from 2 to 7 per cent of unknown mixtures. The following final example illustrates the identification of a fairly complex unknown mixture with a component present in such small amounts that two coded lines do not appear in the pattern.

Example 6—The data in Table 20 were obtained from a sample of primer paint from a piece of industrial machinery. Spectrographic analysis indicated the major metallic elements present to be lead, iron, and zinc. Chromium was found in lower concentration, together with trace amounts of copper, silicon, calcium, magnesium, sodium, bismuth, and titanium. The following would be the steps in a search based on the above recommended procedure.

Step 1—Remove from Deck E the following negative d-value cards, whose intervals include the d-values of the lines of the unknown pattern: Blu/90, 60, 40, 36, 34, 33, 32, 31, 30, 29, 28, 27, 26, 25, 24, 22, 21, 20, 15. Superimpose the remaining 18 negative d-value cards on the viewing box for use in the search.

Step 2—Add the following negative element cards from Deck D to the 18 negative d-value cards selected for use above: Blu/Au, B, Be, Ca, Li, T, RE, Si, Ti, and V. This selection concentrates the search on major components, and neglects trace elements at first. Also note that the cards which include hydrogen, oxygen, nitrogen, phosphorus, sulfur, carbon and the halogens were removed, since these elements are involved in many of the common anions.

Step 3—The superposition of these negative d value and negative element cards leaves several hundred coincident holes. It now becomes desirable to reduce this number further by adding various combinations of positive element and positive d-value cards. Chance contributes considerably to the speed with which an identification is achieved in this step, since it is not always possible to choose combinations that lead to identifications. To avoid the problem of trying to select initially a correct combination of positive element and positive d-value cards, the negative cards for Fe and Pb, Blu/Mn, Sn, were first added together with the positive card for Zn, Wht/Zn, since it was believed that few zinc pigments contain Pb and Fe. This left 29 coincident holes, all of which were checked. As shown in Table 20, zinc oxide (ZnO) and zinc chromate ($ZnCrO_4$) were found to show very good agreement with a number of lines in the unknown, and were tentatively accepted as probable minor constituents of the pigment mixture. Two forms of zinc sulfide (ZnS), wurtzite, and sphalerite, also agreed with several lines, but the intensities would suggest that they constitute only a very small fraction of the mixture, if present at all.

Step 4—It now seems desirable to look for compounds of Pb and Fe to account for the major components of the mixture. The procedure used was the following:

1. Remove the cards Wht/Zn, Blu/Mn, and Blu/Sn from the deck.

2. Add the positive card Wht/Pb and the positive d value card Blk/36 to see if one of the unidentified components is a compound of lead having a high in-

tensity line with $d = 3.39$. This left only 11 coincident holes, and fortunately one of these corresponded to lead oxide (Pb_3O_4) which accounts for about one-half the unidentified lines of the unknown pattern. (Note that, had this choice not led to an identification, the positive card Wht/Fe would have been substituted for Wht/Pb.)

3. The strongest unidentified line is now the one with $d = 2.69$ and $I/I_1 = 100$; therefore, the positive d-value card Ppl/71 corresponding to this line, together with the card Wht/Fe, was added to the deck, and the cards Wht/Pb and Blk/36 were removed. This left 10 coincident holes. One of these corresponded to α-Fe_2O_3 and another to $(Cr,Fe)_2O_3$. Both of these have essentially the same pattern, which accounts satisfactorily for most of the remaining lines of the unknown.

4. There now remained only the three lines having d values of 3.08, 2.94, and 2.75 A that have not been satisfactorily accounted for. All of these are weak and therefore probably belong to one or more phases present in very small amounts. As a first guess, it was assumed that only one phase was involved, and the positive d-value cards for all three lines, Blk/10, Ppl/95, and Ppl/74, were superimposed. This left 10 coincident holes, one of which corresponded to PbO whose three strongest lines matched the unidentified lines.

If this last step had not worked so fortuitously, it would have been necessary to try combinations of one or two of these positive d-value cards with various positive element cards, or with cards corresponding to lines already identified. Alternately it might prove easier to use one of the book indexes to search for appropriate compounds to account for these lines. Examination of Table 20 shows clearly, however, that the use of the negative d-value cards selected in step one would preclude the identification of PbO, because they include card Blu/23, which eliminates all compounds having coded lines in the interval from 2.300 to 2.399, including PbO whose coded line with $d = 2.38$ does not appear in the unknown pattern.

Summary

Two new indexes for use with the Powder Diffraction File in the identification of unknown crystalline materials from their powder diffraction patterns have recently been developed by the editorial board of the Joint Committee for Powder Diffraction Methods. It is believed that these indexes will prove much more convenient and versatile to use for many applications than previous indexes. This paper has reviewed the development, construction, and use of these new indexes in considerable detail to provide a readily available source of detailed information on them and to promote the effective use of these indexes by persons working with the data file. The authors wish to express their sincere appreciation to Karl Beu and J. D. Hanawalt, and to the many other persons who have contributed to the design of the indexes and to the development of procedures for using them.

References

(1) J. D. Hanawalt and H. W. Rinn, "Identification of Crystalline Materials," *Industrial and Engineering Chemistry*, Analytical Edition, Vol. 8, 1936, p. 244.

(2) J. D. Hanawalt, H. W. Rinn, and L. K. Frevel, "Chemical Analysis by X-Ray Diffraction," *Industrial and Engineering Chemistry*, Analytical Edition, Vol. 10, 1938, p. 457.

(3) L. K. Frevel, "Chemical Analysis by Pow-

der Diffraction," *Industrial and Engineering Chemistry*, Analytical Edition, Vol. 16, 1944, p. 209.
(4) L. K. Frevel, "Indexing Powder Patterns of Isomorphous Substances," *Journal of Applied Physics*, Vol. 13, 1942, p. 109.
(5) L. K. Frevel, "Tabulated Diffraction Data for Cubic Isomorphs," *Industrial and Engineering Chemistry*, Analytical Edition, Vol. 14, 1942, p. 687.
(6) L. K. Frevel, H. W. Rinn, and H. C. Anderson, "Tabulated Diffraction Data for Tetragonal Isomorphs," *Industrial and Engineering Chemistry*, Analytical Edition, Vol. 18, 1946, p. 83.
(7) L. K. Frevel and H. W. Rinn, "Tabulated Diffraction Data for Hexagonal Isomorphs," *Analytical Chemistry*, Vol. 25, 1953, p. 1697.
(8) W. L. Fink, "Activities of the Joint Committee on Chemical Analysis by Powder Diffraction Methods, I: X-Ray Data," *Scientific Information in the Fields of Crystallography and Solid State Physics*, Crystallographic Society of Japan, Tokyo, 1962, p. 61.
(9) W. C. Bigelow and Karl E. Beu, "Activities of the Joint Committee on Chemical Analysis by Powder Diffraction Methods, II: A Study of the Problem of Providing Reference Data for Electron Diffraction," *Scientific Information in the Fields of Crystallography and Solid State Physics*, Crystallographic Society of Japan, Tokyo, 1962, p. 99.
(10) F. W. Matthews, "A Coordinate Index to X-Ray Powder Data Using Punched Cards," *Materials Research and Standards*, Vol. 2, 1962, p. 643.

THIS PUBLICATION is one of many issued by the American Society for Testing and Materials in connection with its work of promoting knowledge of the properties of materials and developing standard specifications and tests for materials. Much of the data result from the voluntary contributions of many of the country's leading technical authorities from industry, scientific agencies, and government.

Over the years the Society has published many technical symposiums, reports, and special books. These may consist of a series of technical papers, reports by the ASTM technical committees, or compilations of data developed in special Society groups with many organizations cooperating. A list of ASTM publications and information on the work of the Society will be furnished on request.